ASSOCIATION OF ASIAN STUDIES
MONOGRAPHS AND PAPERS

XI. *Agricultural Involution*

AGRICULTURAL INVOLUTION

Agricultural Involution

THE PROCESS OF ECOLOGICAL CHANGE
IN INDONESIA.   By Clifford Geertz

PUBLISHED FOR THE ASSOCIATION OF ASIAN STUDIES BY

UNIVERSITY OF CALIFORNIA PRESS

Berkeley, Los Angeles, and London

Publication of this volume has been made possible by a generous grant to the Association of Asian Studies by the Ford Foundation.

University of California Press
Berkeley and Los Angeles, California

University of California Press, Ltd.
London, England

© 1963 by The Regents of the University of California

Library of Congress Catalog Card Number 63–20356
Printed in the United States of America

ISBN: 0–520–00459–0 (paper-bound)
      0–520–00458–2 (cloth-bound)

Just as the progress of a disease shows a doctor the secret life of a body, so to the historian the progress of a great calamity yields valuable information about the nature of the society so stricken.

MARC BLOCH

FOREWORD

Interdisciplinary research is always a gamble. All too often it results in a series of discrete articles or monographs with no very clear connecting theme. When the interdisciplinary research is applied to a political or geographic area such as Indonesia, the danger is that the spatial limits will be the only common feature of the products of the different disciplines. Accordingly, when the contributions from various disciplines to the study of a particular area produce genuine synthesis, the result is of unusual interest. Such a synthesis seems to have emerged from the work done on Indonesia under the auspices of the Massachusetts Institute of Technology Center for International Studies.

The Center's research on Indonesia fell into two broad categories: the work of the Indonesia Project, under my direction, which was one of the four major undertakings of the Economic and Political Development Program; and the Indonesia Field Team, under the direction of Rufus Hendon, of which Clifford Geertz, author of the present study, was a member. The field work of both projects was undertaken in the period 1952–1959, and was concentrated mainly in the first half of this period. During the writing-up phase, starting in the second half of this period and only now nearing its end, there has been more or less continuous discussion between the two teams, and particularly between Dr. Geertz and myself.

The M. I. T. Indonesia project was in any case a loosely defined affair, the precise composition of which changed from year to year. It was more a "community of scholars" with a special interest in Indonesia than a tightly organized and centrally directed program of team research. Indeed, I was the only one to be continuously on the payroll throughout the five years of the project's official life. When, however, most of the people who had been associated with the project were gathered together for several days of discussion shortly before the official termination of the project, it became apparent that together we had arrived at an integrated view of Indonesia and its problems. As the various pieces of the picture were laid down, they fitted together like parts of a jig-saw puzzle. Each of us, using the methods of his own discipline—with some tendency in each case to stray over the borderlines of his field—had arrived at essentially the same broad analytical framework, and at the same general conception of the task that lies ahead of the Indonesian people if the high hopes of their revolution are to be realized.

The anthropologist's version of this common picture is set forth more clearly in the present volume than in Dr. Geertz's two earlier works on Indonesia, the first of which deals with religion in Java, whereas the other compares entrepreneurship in Bali with that in east-central Java. The first of these stays closer to the traditional scope and method of anthropology, although the underlying interest in entrepreneurship and economic development is already apparent. This interest is more explicit in the second volume. The present work, however, is nothing less than a social-economic history of Java, that goes far to explain the difficulties that independent Indonesia is experiencing in launching a takeoff into sustained economic growth, after three centuries of apparently static living levels under Dutch colonialism.

It may help the reader to put Dr. Geertz's latest volume on

Indonesia into its broader context if a brief outline of the common thesis—common to the economists, political scientists, historians, sociologists, anthropologists, and geographers associated with our Indonesian venture—is provided here by way of introduction. The language is my own, and I would not wish to commit any other member of the group to it; but I have tried to render as faithfully as possible the common viewpoint expressed on those occasions when we have met together.

In Indonesia a "big push" is needed, not only from the economic point of view, but from the sociological and political viewpoints as well. There seems to be in the history of each country an "optimal moment" for launching development, a short period of time when sociological, political, and economic factors coalesce to provide a climate unusually favorable for a take-off into economic growth. If such an optimal moment is missed, it may take several generations to reproduce a similar set of favorable conditions. In simplest form the thesis sounds a bit tautological: a take-off is most likely when the growth of an indigenous entrepreneurial class occurs simultaneously with the appearance of a political elite which has the power and the will to provide a policy framework favorable for the exercise of entrepreneurial talents—whether in the public or the private sector. Yet as a guide to historical analysis—and even for prognosis—the thesis seems to us a useful one.

The story of Java seems to be one of repeated nipping off of a budding entrepreneurial upsurge by a political elite essentially hostile to it. It may even be that Javanese trade was prevented from expanding by Portuguese enterprise. Certainly any hopes the Mataram Empire may have had of reuniting Indonesia—to the degree, say, that the country was united by the Madjapahit Empire two centuries earlier—were destroyed through use of main force by the Netherlands East India Company.

About 1850, under the impetus of the Culture System and the first wave of Dutch settlement and plantation enterprise, another incipient "take-off" emerged. The Dutch settlement also launched a population explosion; but there is reason to believe that, had the new class of Javanese entrepreneurs been given their head, they would have brought increases in productivity and changes in social structure and attitudes that would have permitted a continued rise in the per-capita income nonetheless. Unfortunately, the Dutch colonial administrators were the effective political elite at the time, and Dutch colonial policy was not directed toward the encouragement of any such form of development. There is some evidence—not too clear—that the great depression of the 1930's, by bringing Dutch enterprise to a standstill, provided a last injection of encouragement to Javanese industry and rubber exporters. On the other hand, the depression killed off whatever remained of Javanese enterprise in sugar. Such potential as Javanese society still had for generating the conjuncture of economic and political leadership needed for a take-off was subsequently destroyed by the war, the Japanese occupation, the revolution, and the subsequent political chaos.

Meanwhile—for all their "hands-off" policy—the intensity of Dutch economic activity in Java detraditionalized Javanese society. When Javanese aristocrats, thwarted in their efforts to extend their leadership into the economic field, retreated into Javanese tradition—except for a handful with Western training who sought the political sphere as an outlet for their talents and ambitions—they were pursuing a shibboleth. Traditional Javanese society no longer existed. It is for this reason that the only leadership capable of arousing a *mass* following in Java in recent years have been of persuasions—Communist or extreme nationalist—having little or nothing to do with Javanese tradition. But while Communists and nationalists provide political leadership, they do not at this stage provide economic leadership or entrepreneurship—if anything, their actions and attitudes are such as to discourage the growth of entrepreneurial activity.

In the rest of Indonesia the impact of the Dutch came much later, was limited to a few decades, and was neither so wide nor so deep in its effects as it was in Java. The societies of North

Sumatra (Batak), Central Sumatra (Minangkebau), Bali, Sulawesi, and Kalimantan (Dajak) were much less detraditionalized
by contact with the Dutch. Consequently, they now have a chance
to evolve in terms of their own cultural dynamics, stimulated by
the liberating effect of the revolution. Each of these societies has
developed a group of indigenous entrepreneurs, some of whom—
especially Batak and Minangkebau—are providing economic
leadership to Java as well as to their own regions. Moreover,
these societies—again, especially Batak and Minangkebau—are
providing political leadership far out of proportion to their numbers, on the national scene as well as at home. In the Outer
Islands there is still reason to hope for the kind of interaction of
entrepreneur and elite that could give rise to sustained economic
growth. It is not just that the Outer Islands have a much more
favorable resource-population pattern than Java, although that is
also true; it is the Outer Islands that constitute the "growing
points and leading sectors" in sociological terms as well. Here is
part of the explanation of the "dualism" so often noted in the
Indonesian economy.

This "dualism" resulted in large measure from the special
form of industrialization by foreign enterprise after the Napoleonic wars. Until about 1820 the entire country was given over to
small holders' and peasant agriculture, with no marked difference
in technique from one sector or region to another. The population of Indonesia was small—less than 10 percent of its present
level, with less than 5 million people on Java and perhaps 3 million in the Outer Islands. Land was therefore relatively abundant.
There was no serious population pressure anywhere in the country and, while standards of living were not high, there was little
real hardship. The Dutch were for the most part content to trade
with the Indonesians (to a large extent through Chinese middle-
men) and later to obtain forced deliveries of export products,
without themselves settling in the country.

After the brief interlude of British rule during the Napoleonic
wars, the Dutch colonial administration in Indonesia gradually

shifted to plantation agriculture. Some decades later the development of the mining, petroleum, and processing industries began.
This industrialization seems to have brought an initial increase
in *per capita* incomes of Indonesians. However, instead of leading to permanent improvement in Indonesian living standards,
this initial increase in incomes was rapidly offset by accelerated
population growth. The higher incomes afforded subsistence for
greater numbers. Moreover, the shift in policy brought Indonesians in closer contact with Western civilization: industrialization meant settlement by Europeans. Efforts were made to raise
levels of health and to maintain peace and order among the Indonesians who were previously rather given to fighting each
other. Death rates dropped. It may also be—although this is less
clear—that improved health and nutrition raised fertility rates.
The form of development, centered on plantations, mines, and
oil fields, all of which produced raw materials for export, brought
more *industrialization* than *urbanization;* the checks on family
size brought by urban industrialization of Europe and the New
World were therefore less effective in Indonesia. The net effect
was that the total population increased more than fourfold in
three generations.

The sector of the economy in which new investments were being made was totally incapable of absorbing the increase in population which it generated, because the plantation, industry, oil and
mining operations were land- and capital-intensive; technical coefficients either were, or were assumed to be, relatively fixed. The
increased numbers inevitably led to a return flow into peasant
agriculture and small cottage industries, where technical coefficients were relatively variable and opportunities for obtaining a
living still existed within the village structure. With abundant
labor and scarce capital—and in Java scarce land as well—production methods in this sector became highly labor-intensive.
(Even in the Outer Islands, there is now no obvious superabun-
dance of fertile, easily cleared, readily cultivatable land. And in
recent decades population growth in the Outer Islands has been

nearly as rapid as it was in Java during the late nineteenth century.) Eventually labor became redundant in this sector (marginal productivity fell to zero) and the growing population led merely to work spreading, underemployment and unemployment. *Per capita* incomes in this sector returned to the subsistence level.

Technological progress was confined largely to the capital-intensive sector. In the labor-intensive sector, meanwhile, the rate of population growth far exceeded capital accumulation. More recently, trade union activity and government policy have established industrial wage levels which, low as they are, are sometimes high in relation to the marginal productivity of labor. This situation further aggravates the tendency to use labor-saving devices wherever possible in the industrial sector. On the other hand, there is no incentive for groups of individual farmers or small enterprises to introduce labor-saving but capital-absorbing innovations in the rural sector. A technology that will raise man-hour productivity without net investment has yet to be discovered. Nor is there any incentive for labor *as a group* to increase its efforts, since the labor supply is already redundant. Thus methods remain labor-intensive, and levels of technique, man-hour productivity, and economic and social welfare remain low in the peasant-agriculture and small-industry sector.

Both the economic and the political aspects of this problem are made more difficult of solution by the fact that the two sectors conform roughly to two regions: Java and the Outer Islands. The main development of plantations, mines, and oil fields took place on the Outer Islands, especially Sumatra, Kalimantan, and Sulawesi. The big growth of population, however, took place where the soil was most fertile and best suited for growing foodstuffs, and where most of the people already were—Java. Employment in cottage and small industry is also highest in Java.

Here then is a country in which two-thirds of the population are crowded on one relatively small island, engaged mainly in the production of foodstuffs or simple handicrafts and small-scale manufacturing for home consumption, and dependent on imports for textiles and other essential items of consumption; while the other third of the population is scattered through an enormous area, in which there are some highly efficient large-scale industries producing mainly for export. With such a discrepancy in current economic conditions, as well as in capacity for generating leadership for economic growth, it is small wonder that there are stresses and strains between Java and the Outer Islands.

The implications of this analysis are far-reaching indeed. For one thing, it means that President Sukarno was responding to a very real problem in demanding constitutional reform in the direction of "democracy with leadership." At this stage ordinary parliamentary democracy must mean essentially rule by Java, since Java has two-thirds of the electorate. The only way in which Java could now provide the required coalescence of entrepreneurs and elite would be through growth of the Communist party and the establishment of a Communist regime. Yet the leaders of the Revolutionary Government (P.R.R.I.) were also right; maximizing the growth potential of the Indonesian economy means not only allocating more of the development budget to public or private projects in the Outer Islands, it also means allocating much more responsibility for financing and executing development to the regions.

The current Eight Year Plan falls short of requirements cast in such terms. It seeks to crystallize the conception of Indonesian democracy and to define the sort of society the plan is expected to produce. The spirit of the plan is well indicated by its division into 8 volumes, 17 parts, and 1,945 paragraphs, to symbolize the Proclamation of Independence on August 17, 1945. The Indonesian national identity, the plan states, is to be sought in the traditional culture—the songs, the dance, the *wayang,* the literature. The goal must be "a just and prosperous society based on the Pantja Sila," the five pillars of Indonesian social and political philosophy—nationalism, humanitarianism, democracy, social justice, and belief in God. It must also be a "family-like society," reproducing at the national level the spirit of village organization.

The definition of "socialism à la Indonesia" is useful; but the plan does not really tackle the basic economic problem. It does *not* impose any "minimum effort" on the government. The plan for the central government—and there is no authorized and detailed plan for the regional and local governments or for private enterprise—is essentially a projection of programs already under way in the various departments. The plan does not require of the government anything essentially *new,* let alone bold or imaginative, in the way of developmental effort. It does not really bind the government to actions constituting a clean break from what they have been doing in the past. Consequently, it does not enforce on the government any of the basic decisions on economic policy that must be made before development can proceed. On the contrary, it permits the government to go on postponing really serious consideration of the development problem, while stating to themselves and to the people that they are "carrying out a development plan."

Indonesian development will be expensive. Yet the events of 1957 and 1958, together with stagnation of real income and mounting inflation, make it abundantly clear that if Indonesia is to become a genuinely unified and prosperous nation, a take-off into visible economic growth cannot be delayed much longer.

The more purely economic aspect of Indonesia's history and current problems will be analyzed in the final volume of the M. I. T. project. Dr. Geertz's present study provides the sociocultural setting and makes the economist's task much easier.

BENJAMIN HIGGINS

Berkeley, California, May, 1963

ACKNOWLEDGMENTS

This monograph is an expansion and comprehensive revision of the first section of my earlier programmatic essay, "The Development of the Javanese Economy: A Socio-Cultural Approach," issued in dittoed form by the Center for International Studies of the Massachusetts Institute of Technology in 1956 as document C/56-1 in their Economic Development Program series. I am most grateful to the Center and to its director, Max Millikan, for supporting both the writing of that original report and the anthropological fieldwork I carried out in Java, as part of the "Modjokuto Project," in 1952–1954, in the course of which most of the ideas and interpretations here worked out in historical terms were originally developed.

The writing of the present study has been made possible by the Committee for the Comparative Study of New Nations of the University of Chicago, and I am grateful to the Committee as a whole, to its chairman, Edward Shils, and to its individual members for providing this opportunity. Among those who have read and criticized various drafts of this work are: Robert M. Adams, Harold Conklin, Lloyd A. Fallers, Hildred Geertz, Benjamin Higgins, Robert Jay, Morris Janowitz, Kampto Utomo,

W. F. Wertheim, and Aram Yengoyan, and I wish to express my appreciation for their assistance. Messrs. Conklin, Higgins, Kampto Utomo, and Yengoyan, in particular, provided extensive and detailed comments and criticism, most of which I have in some way incorporated into the work, though they are, of course, not to be held responsible for my interpretations of their arguments. I am indebted to Donald McVicker for rendering the charts and maps.

In most general terms, this book is an attempt to apply to the interpretation of—in this case, economic—history some concepts and findings of social anthropology, to utilize the insights derived from microsociological analysis for understanding macrosociological problems, and to establish a fruitful interaction between the biological, social, and historical sciences. However well or poorly this particular work effects such an integration of different scholarly perspectives, I am convinced that an adequate understanding of the new countries of the "third world" demands that one pursue scientific quarry across any fenced-off academic fields into which it may happen to wander. And whatever substantive merits my analysis may have, I hope that my effort will at least suggest the profits to be gained in such poaching expeditions.

C. G.

Chicago, January, 1963.

CONTENTS

III: THE OUTCOME

PART I

Starting Points, Theoretical and Factual

1. THE ECOLOGICAL APPROACH IN ANTHROPOLOGY

The recent burst of efforts to adapt the biological discipline of ecology—the science which deals with the functional relationships between organisms and their environment—to the study of man is not simply one more expression of the common ambition of social scientists to disguise themselves as "real scientists," nor is it a mere fad. The necessity of seeing man against the well-outlined background of his habitat is an old, ineradicable theme in anthropology, a fundamental premise. But until recently this premise worked out in practice in one of two unsatisfying forms, "anthropogeography" or "possibilism"; and the turn to ecology represents a search for a more penetrating frame of analysis within which to study the interaction of man with the rest of nature than either of these provides.

The Limitations of Traditional Approaches

In the anthropogeographic approach, of which the climatological theories of Elsworth Huntington are the most famous, if hardly the most sophisticated, example, the problem was phrased in terms of an investigation of the degree and manner in which human culture was shaped by environmental conditions.[1] This position did not necessarily involve a thorough-going environmental determinism, because some variation in human culture independent of geographic forces was admitted by even the most extreme members of this school. But such variations were put down to "accident," the escape hatch of ethnology, or, on occasion, to "race," the escape hatch of biology. In the possibilist approach, on the other hand, the environment was seen not as causative but as merely limiting or selective. Geographical factors did not *shape* human culture—a wholly historical, even "superorganic" phenomenon—but they set boundaries to the forms it could take at any place and time. A. L. Kroeber's classic discussion of the confinement of maize-growing in aboriginal North America to regions where a 120-day period with sufficient rain and without killing frosts existed, is an example of this even more popular type of analysis: the nature of the environment does nothing in itself to stimulate the growing of maize, but it can insure the nongrowing of it.[2]

Neither of these views is simply wrong, yet both are inadequate for precise analysis. Geographic factors do often seem to play, as the anthropogeographers argued against the possibilists, a dynamic, not merely a passive role in the development of human culture. But at the same time the direct derivation of virtually any specific cultural practice from the nature of the geographical habitat as such seems to be, as the possibilists argued against the anthropogeographers, a nonsequitur: maize-farming may have been well adapted to the physical conditions of the pre-Columbian Southwest (or those conditions to it), but it can hardly be said to have been caused by them. The indeterminacy on either side here actually stems from a serious conceptual defect the two approaches share. Both initially separate the works of man and the processes of nature into different spheres—"culture" and "environment"—and then attempt subsequently to see how

---

[1] Huntington, 1945. See also, Semple, 1911.

[2] Kroeber, 1939, pp. 207–212. For other statements of the possibilist position, see Forde, 1948; and Wissler, 1926.

as independent wholes these externally related spheres affect one another. With such a formulation, one can ask only the grossest of questions: "How far is culture influenced by environment?" "How far is the environment modified by the activities of man?" And can give only the grossest of answers: "To a degree, but not completely."

The ecological approach attempts to achieve a more exact specification of the relations between selected human activities, biological transactions, and physical processes by including them within a single analytical system, an *ecosystem*. In ecology generally, an ecosystem consists of a biotic community of interrelated organisms together with their common habitat and can range in size, scope, and durability from a drop of pond water together with the micro-organisms which live within it to the entire earth with all of its plant and animal inhabitants.[3] The concept of an ecosystem thus emphasizes the material interdependencies among the group of organisms which form a community and the relevant physical features of the setting in which they are found, and the scientific task becomes one of investigating the internal dynamics of such systems and the ways in which they develop and change. "When the ecologist enters a field or meadow," Paul Sears has written, "he sees not what is there but what is happening there." [4]

What is happening there is a patterned interchange of energy among the various components of the ecosystem as living things take in material as food from their surroundings and discharge material back into those surroundings as waste products, a process Haeckel, who was perhaps the founder of the field of ecology (at least he coined its name), aptly called "external physiology." [5] And as in internal physiology, so in external, the maintenance

---

[3] Dice, 1955, p. 2.

[4] Sears, 1939, quoted in Clarke, 1954, p. 16.

[5] "Just as morphology falls into two main divisions of anatomy and development, so physiology may be divided into a study of inner and outer phenomena. . . . The first is concerned with the functioning of the organ-

of system equilibrium or homeostasis is the central organizing force, commonly referred to in this context as "the balance of nature." [6] If one takes, for example, a flock of sheep in a pasture, the sheep are, with their sharp, close-cropping teeth, apparently destroying the grass by ingesting it. But the sheep are also fertilizing the pasture with their manure. Thus, if the sheep were removed the pasture would, at least in many cases, be removed too; for trees would begin to seed and grow, finally killing off the pasture grass, and where once was a field would now be a wood. The sheep and the pasture form an integrated, equilibrated system, each of them dependent upon the other for its existence. Such equilibria are commonly, of course, quite complex—consider the neat balance between water, oxygen, light, heat, green plants, microscopic animals, insects, and fishes in a pond.

Nor does the inclusion of man as an element in an ecosystem change the nature of the basic principles. Clarke, from whom the sheep-in-the-pasture example is drawn, tells of ranchers who, disturbed by losses of young sheep to coyotes, slaughtered, through collective effort, nearly all coyotes in the immediate area. Following the removal of coyotes, the rabbits, field mice, and other small rodents, upon whom the coyotes had previously preyed, multiplied rapidly and made serious inroads on the grass of the pastures. When this was realized, the sheep men ceased to kill coyotes and instituted an elaborate program for the poisoning of rodents. The coyotes filtered in from the surrounding areas, but finding their natural rodent food now scarce, were forced to

---

ism in itself, the second with its relationships with the outer world. . . . By *ecology*, we understand the study of the economy, of the household, of animal organisms. This includes the relationships of animals with both the organic and inorganic environments, above all the beneficial and inimical relations with other animals and plants, whether direct or indirect." Haeckel, 1870; quoted in Bates, 1953.

[6] Odum, 1959, p. 25. The most systematic treatment of homeostasis as a general phenomenon is Ashby, 1960.

turn with even greater intensity to the young sheep as their only available source of food.[7]

Nevertheless, the adaptation of the principles of ecological analysis and of the concepts in terms of which they are expressed (niche, succession, climax, food chain, commensality, trophic level, productivity, and so on) to the study of man can be conducted in a variety of manners, not all of which are equally useful.[8] The simplest method is merely to view the whole of human society as basically a biotic phenomenon like any other and to apply ecological concepts to it directly and comprehensively, an approach characteristic of the school of "urban," "social," or "human" ecology founded by the sociologist Robert Park.[9] In practice, most of such analyses turn out to be investigations in what might be more properly called "locational theory" than ecology. Not only are the biological concepts employed more analogically than literally, but a fundamentally a-cultural view of human society is adopted which sees settlement patterns, and in fact human activities generally, as an inevitable result of the free play of competitive "natural" (or "economic") forces, regulated, save for slight and temporary distortions introduced by customs, sentiments and values, by the principle of least costs. In any case, this reductionist use of ecology as an exclusive and comprehensive frame for the analysis of human community structure is not intended here. When we speak of ecological analysis we are concerned not with "explaining the territorial arrangements that social activities assume . . . the regularities which appear in man's adaptation to space," [10] but with determining the

---

[7] Clarke, 1954, p. 19.

[8] Bates, 1953, offers a survey of the divergent ways in which ecology has been used as a label for human studies, some of which amount to hardly more than sloganeering.

[9] Park, 1934, 1936. For more recent formulations, see Hawley, 1950; and Quinn, 1950. For a brilliant and devastating critique of his whole approach, see Firey, 1947, pp. 3–38.

[10] Firey, 1947, p. 3.

relationships which obtain between the processes of external physiology in which man is, in the nature of things, inextricably embedded, and the social and cultural processes in which he is, with equal inextricability, also embedded.

Cultural Ecology

Much closer to the perspective adopted here is that of Julian Steward, who has been developing a mode of analysis he calls "cultural ecology." [11] The distinctive feature of his approach is a strict confinement of the application of ecological principles and concepts to explicitly delimited aspects of human social and cultural life for which they are particularly appropriate rather than extending them, broadly and grandly, to the whole of it. The still powerful anthropological doctrine of "holism," which holds all aspects of culture to be fully interdependent, leads to a formulation of the culture-environment problem in gross overall terms and thus to the "there is something in both arguments" paradox already mentioned. Generally characterized habitat types —"the tropics," "the polar regions," "the high plains"—are matched to whole, and presumptively integral, cultures—"the Javanese," "the Eskimo," "the Sioux." On such a global level, Huntington, for all his simplistic excesses, can make a case that climate does somehow affect culture, for surely there is something vaguely arctic about the Eskimo, tropical about the Javanese. But Hegel can, with equal plausibility, dismiss environmental determinism with the fine Johnsonian argument that "where the Greeks once lived, the Turks now live; and that's an end on the matter."

Steward, however, rather than asserting that all aspects of culture are, in some indeterminate way, functionally interrelated, argues that the degree and kind of interrelationship is not the same in all aspects of culture, but varies. He attempts to isolate in the culture he analyzes certain aspects in which functional

[11] Steward, 1955, pp. 30–42.

ties with the natural setting are most explicit, in which the interdependency between cultural patterns and organism-environment relationship is most apparent and most crucial. These aspects of the wider culture he terms the "cultural core," while to aspects not so closely related to adaptive processes he merely refers, rather lamely, as "the rest of culture." And it is to the core alone that ecological analysis is relevant:

[The cultural core refers to] the constellation of features which are most closely related to subsistence activities and economic arrangements. The core includes such social, political and religious patterns as are empirically determined to be closely connected with these arrangements. Innumerable other features may have greater potential variability because they are less strongly tied to the core. These latter, or secondary features, are determined to a greater extent by purely cultural-historical factors—by random innovations or by diffusion—and they give the appearance of outward distinctiveness to cultures with similar cores. Cultural ecology pays primary attention to those features which empirical analysis shows to be most closely involved in the utilization of environment in culturally prescribed ways.[12]

A correlative analysis of the environmental side of the equation is also undertaken. It is reduced from a gross variable more or less identical to the whole habitat, geographically considered, to those selected features which actually have functional significance for human adaptation in any given case. Steward points out, for example, that noncultivating societies with essentially the same hunting technology (bow, spears, deadfalls) may differ in various ways as a result of the kind of animals which exist in their environment. If the main game animal exists in large herds, say, bison or caribou, it is adaptive to engage in cooperative hunting on a fairly sizable scale. Considerable numbers of people are likely to remain together throughout the year, following the herds as they move, driving them in mass surrounds, and so

---

[12] Steward, 1955, p. 37.

on. If, however, the game is of the sort which occurs in small scattered groups and does not migrate, it is better hunted piecemeal by small groups of men who know their immediate territory extremely well—large population concentrations being impossible at any rate. In the first situation, Steward argues, the elementary community will tend to be a relatively large, multifamily group, while in the second it will tend to be a small, localized patrilineal band. These cross-cultural organizational similarities occur, in either situation, not because of total habitat similarity, but because crucial elements in the environment—the type and distribution of game—are similar. Thus, the patrilineal-band—small-animal situation is found among the Bushmen, who live in a desert, the Negritos, who live in rain forests, and the Fuegians, who live on a cold, rainy littoral plain. These groups show similar social structural features despite this contrast in habitats, because their environments are similar in the important matter, for hunting peoples, of the type of game they contain.[13]

These two exercises in the disaggregation of global variables, the discrimination of the "cultural core," and the definition of the relevant environment, are directly reciprocal endeavors. If one empirically determines the constellation of cultural features which are most unequivocally related to the processes of energy interchange between man and his surroundings in any given instance, one necessarily also determines which environmental features have primary relevance for those same processes. The sharpness of the division between analyses from the side of "man" and analyses from the side of "nature" therefore disappears, for the two approaches are essentially alternative and interchangeable conceptualizations of the same systemic process. Ashby has formulated this fundamental principle in more general terms:

As the organism and its environment are to be treated as a single system, the dividing line between "organism" and "environment" becomes partly conceptual, and to that extent arbitrary. Anatomically and physically, of course, there is usually a unique and obvious dis-

---

[13] Steward, 1955, pp. 122–50.

tinction between the two parts of the system; but if we view the system functionally, ignoring purely anatomical facts as irrelevant, the division of the system into "organism" and "environment" becomes vague. Thus, if a mechanic with an artificial arm is trying to repair an engine, then the arm may be regarded either as part of the organism that is struggling with the engine, or as part of the machinery with which man is struggling . . . The chisel in a sculptor's hand can be regarded either as part of the complex biophysical mechanism that is shaping the marble, or it can be regarded as part of the material which the nervous system is attempting to control.[14]

On a more explicitly cultural level the situation is similar. The Eskimo's igloo can be seen as a most important cultural weapon in his resourceful struggle against the arctic climate, or it can be seen as a, to him, highly relevant feature of the physical landscape within which he is set and in terms of which he must adapt. A Javanese peasant's terrace, to use a more directly pertinent example, is both a product of an extended historical process of cultural development and perhaps the most immediately significant constituent of his "natural" environment.

Nor, again, are only elements of so-called "material culture" conceivable in such terms. Intimately connected with the igloo are Eskimo settlement patterns, family organization, and sexual division of labor. Javanese rice terraces are closely integrated with modes of work organization, forms of village structure, and processes of social stratification. As one specifies more fully the precise nature of a people's adaptation from the geographical side, one inescapably specifies, at the same time and to the same degree, their adaptation from the cultural side, and vice versa. One delineates, in short, an ecosystem within which certain selected cultural, biological, and physical variables are determinately interrelated, and which will yield to the same general mode of analysis as ecosystems within which human organisms do not happen to play a role.

---

[14] Ashby, 1960, p. 40.

This mode of analysis is of a sort which trains attention on the pervasive properties of systems *qua* systems (system structure, system equilibrium, system change) rather than on the point-to-point relationships between paired variables of the "culture" and "nature" variety. The guiding question shifts from: "Do habitat conditions (partly or completely) cause culture or do they merely limit it?" to such more incisive queries as: "Given an ecosystem defined through the parallel discrimination of cultural core and relevant environment, how is it organized?" "What are the mechanisms which regulate its functioning?" "What degree and type of stability does it have?" "What is its characteristic line of development and decline?" "How does it compare in these matters with other such systems?" And so on. One conceives of the techniques of swidden agriculture as an integral part of a larger whole which includes alike the edaphic and climatological characteristics of tropical forest landscapes, the social organization of a labor force which must be shifted continually from field to field, and the empirical and nonempirical beliefs which influence the utilization of scattered and varied land resources. Consideration of wet-rice terracing widens out into the complex dynamics of a sort of self-sustaining aquarium on the one hand, and into questions of demography, underemployment, and moral valuations with respect to cooperative endeavor on the other. Yet such systems are bounded; they do not include everything. And, so bounded, the processes by which they develop, maintain their identity, transform themselves or deteriorate, can be specified— as can the influence of the external, parametric conditions which most significantly play upon them. Cultural ecology, like ecology generally, forms an explicitly delimited field of inquiry, not a comprehensive master science.

Yet it is necessary also explicitly to dissent from Steward's apparent assumption that although cultural ecology is not a comprehensive science it is nevertheless a privileged one. Referring to that part of culture "most closely related to subsistence activities and economic arrangements" as the "core" of it, while denoting the rest of culture as "secondary," indeterminately shaped by the accidents of random innovation and diffusion, means begging the question. There is no *a priori* reason why the adaptive realities a given sociocultural system faces have greater or lesser control over its general pattern of development than various other realities with which it is also faced. The best that can be said for such a statement as "over the millennia cultures in different environments have changed tremendously, and these changes are basically traceable to new adaptations required by changing technology and productive arrangements" is that it brings what is elsewhere in Steward's work a *petitio principii* out in the open as a mere prejudice.[15] It is a commendably ambitious proposition, but one which needs proof, not mere assertion. How much of the past growth and present state of Indonesian culture and society is attributable to ecological processes is something to be determined, if at all, at the end of inquiry, not at the beginning of it. And as political, stratificatory, commercial, and intellectual developments, at least, seem to have acted as important ordering processes in Indonesian history, the final awarding of prepotency to ecological developments seems no more likely than that they will turn out to have been inconsequential.

---

[15] Steward, 1955, p. 15.

2. TWO TYPES OF ECOSYSTEMS

Inner vs. Outer Indonesia

A handful of mere statistics of the most routine, humdrum sort can sketch a picture of the basic characteristics of the Indonesian archipelago as a human habitat with more immediacy than pages of vivid prose about steaming volcanoes, serpentine river basins, and still, dark jungles. The land area of the country amounts to about one and one-half million square kilometers, or about that of Alaska. Of this only about one hundred and thirty-two thousand square kilometers are in Java, the rest making up what are usually called "the Outer Islands"—Sumatra, Borneo (Kalimantan), Celebes (Sulawesi), the Moluccas, and the Lesser Sundas (Nusa Tenggara). But the country's total population (1961) is around ninety-seven million, while Java's population alone is about sixty-three million. That is to say, about 9 percent of the land area supports nearly two-thirds of the population; or, reciprocally, more than 90 percent of the land area supports approximately one-third of the population. Put in density terms, Indonesia as a whole has about 60 persons per square kilometer; Java has 480, and the more crowded areas of the central and east-central parts of the island more than a thousand. On the other hand, the whole of Indonesia minus Java (i.e., the Outer Islands) has a density of around twenty-four per square kilometer. To summarize: all over, 60; the Outer Islands, 24; Java, 480: if ever there was a tail which wagged a dog, Java is the tail, Indonesia the dog.[1]

The same plenum and vacuum pattern of contrast between Java and the Outer Islands appears in land utilization. Almost 70 percent of Java is cultivated yearly—one of the highest proportions of cropland to total area of any extensive region in the world—but only about 4 percent of the Outer Islands. Estate agriculture aside, of the minute part of the Outer Islands which is cultivated, about 90 percent is farmed by what is variously known as swidden agriculture, shifting cultivation, or slash-and-burn farming, in which fields are cleared, farmed for one or more years, and then allowed to return to bush for fallowing, usually eventually to be recultivated. On Java, where nearly half the smallholder's crop area is under irrigation, virtually no swidden agriculture remains. In the irrigated regions, field land is in wet-rice terraces, about half of them double-cropped, either with more wet rice or with one or several secondary dry crops. In the unirrigated regions, these dry crops (maize, cassava, sweet potatoes, peanuts, dry rice, vegetables, and others) are grown in a crop-and-fallow regime. Production statistics present, of course, the same picture: in 1956 approximately 63 percent of Indonesia's rice, 74 percent of her maize, 70 percent of her cassava, 60 percent of her sweet potatoes, 86 percent of her peanuts, and 90 percent of her soya beans were produced in Java.[2]

Actually, this fundamental axis of ecological contrast in Indo-

---

[1] Sumaniwata, 1962. Madura is included with Java in the calculations but the transitional area of West New Guinea (Irian) is not included. For a useful general summary of Indonesian demographic realities, see The Population of Indonesia, 1956.

[2] Metcalf, 1952. Statistical Pocketbook of Indonesia, 1957, p. 51. Commercial crop cultivation shows, however, a sharply contrasting pattern.

nesia is not altogether accurately demarcated when one phrases it, following the received practice of the census takers, simply in terms of Java (and Madura) versus the Outer Islands, because in fact the "Javanese" pattern is found in southern Bali and western Lombok as well, and is but weakly represented in the southwestern corner of Java (South Bantam and South Priangan) where a pattern more like that of the Outer Islands, including a certain amount of swidden, is found. Thus, we might better refer to the contrast as one between "Inner Indonesia"—northwest, central, and east Java, south Bali, and west Lombok; and "Outer Indonesia"—the rest of the Outer Islands plus southwest Java, which do in fact form more or less of an arc pivoted on central Java. (See Map 1.) Such a division is, in any case, a gross one which needs modification in detail: patches of relatively intensive irrigation agriculture are found at either tip,

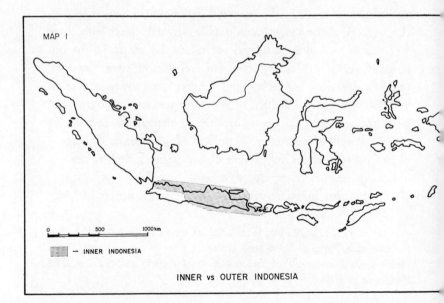

MAP 1

0   500   1000 km

▓▓ — INNER INDONESIA

INNER vs OUTER INDONESIA

around Lake Toba, and in the western highlands in Sumatra as well as in the southwest arm of the Celebes, for example, and eastern Madura deviates somewhat from the Javanese norm.[3] But it does lead, in a broad and general way, to a fruitful discrimination of two different sorts of ecosystems with two different sorts of dynamics—one centering on swidden agriculture, one on wet-rice agriculture—in terms of which the striking differences in population density, modes of land use, and agricultural productivity can be understood.

Swidden

As Conklin has pointed out, much of the inadequate treatment swidden agriculture has received in the literature is a result of the fact that characterizations of it have tended to be negatively phrased.[4] Thus, Gourou outlines as its four most distinctive features: (1) it is practiced on very poor tropical soils; (2) it represents an elementary agricultural technique which utilizes no tool except the axe; (3) it is marked by a low density of population; and (4) it involves a low level of consumption.[5] Similarly, Pelzer says that it is marked by a lack of tillage, less labor input than other methods of cultivation, the nonutilization of draft animals and manuring, and the absence of a concept of private landownership.[6] For Dobby, it represents "a special stage in the evolution from hunting and food gathering to sedentary farming," this specialness evidently consisting of such null traits as nonrelation to pastoral pursuits and the production of very little which is of trading or commercial significance.[7] And for many, by far its most outstanding feature is that singled out by Spate—namely, that its practice is "attended by serious defores-

---

[3] Terra, 1958.
[4] Conklin, 1957, p. 149.
[5] Gourou, 1956.
[6] Pelzer, 1945, pp. 16 ff.
[7] Dobby, 1954, pp. 347–349.

tation and soil erosion." [8] Aside from the fact that most of these depreciatory statements are dubious as unqualified generalizations (and a few are simply incorrect), they are not of much help in understanding how swidden farming systems work.

In ecological terms, the most distinctive positive characteristic of swidden agriculture (and the characteristic most in contrast to wet-rice agriculture) is that it is integrated into and, when genuinely adaptive, maintains the general structure of the pre-existing natural ecosystem into which it is projected, rather than creating and sustaining one organized along novel lines and displaying novel dynamics. In the tropics, to which, for reasons we may postpone considering, this form of cultivation is today largely confined, the systemic congruity between the biotic community man artificially establishes on his swidden plot and that which exists there in stable climax independent of his interference (in the main, some variety of tropical forest) is striking. Any form of agriculture represents an effort to alter a given ecosystem in such a way as to increase the flow of energy to man: but a wet-rice terrace accomplishes this through a bold reworking of the natural landscape; a swidden through a canny imitation of it.

The first systemic characteristic in which a swidden plot simulates a tropical forest is in degree of generalization. By a generalized ecosystem is meant one in which a great variety of species exists, so that the energy produced by the system is distributed among a relatively large number of different species, each of which is represented by a relatively small number of individuals. If, on the contrary, the system is one with a relatively small number of species, each of which is represented by a relatively large number of individuals, it is said to be specialized. Put somewhat more technically, if the ratio between number of species and number of organisms in a biotic community is called its *diversity index,* then a generalized ecosystem is one characterized by a

---

[8] Spate, 1945, p. 527, quoted in Leach, 1954, p. 22.

community with a high diversity index, a specialized one by a ✓
community with a low diversity index. Natural communities
tend to vary widely in their degree of generalization, or the size
of their diversity index: a tropical forest, and in particular a rain
forest, is a very generalized, very diverse community, with an
enormous variety of plant and animal species sporadically represented; a tundra is characterized by a very specialized, uniform
community, with relatively few species but, at least in the sub-
arctic, a large number of clustered individuals.[9]

Much of the most effective human utilization of the natural
habitat consists of changing generalized communities into more
specialized ones, as when natural ponds containing a wide variety
of green plants, aquatic animals, and fishes are transformed into
managed ones in which the number of types of primary plant
producers is sharply reduced to those which will support a few
select types of fish edible by man. The rice terrace, which can, in
these terms, be viewed as a sort of slowly drained, managed pond
focused on an edible plant, is an outstanding example of artificially created specialization. The reverse process, increased generalization, also occurs, of course, as when man introduces into
a temperate grassland area (for example, the American prairie) a
wide variety of interrelated domestic plants and animals, which,
though they constitute a much more diverse community than that
indigenous to the area, nonetheless prove to be viable within it.

Still other human adaptations, however, attempt to utilize the
habitat not through altering its diversity index, but through more
or less maintaining its over-all pattern of composition while
changing selected items of its content; that is, by substituting
certain humanly preferred species for others in functional roles
("niches") within the pre-existing biotic community. This is not
to say that such adaptations do not seriously alter the indigenous
ecosystem (as, in a gross sense, most hunting and gathering adaptations do not), or that their general effect on the balance of

---

[9] These concepts are taken from Odum, 1959, pp. ii, 50–51, 77, 281–283,
316, and 435–437.

nature may not sometimes be a radical one; but merely that they alter the indigenous ecosystem by seeking to replace it with a system which, although some of its concrete elements are different, is similar to it in form, rather than by a system significantly more specialized or more generalized. Large-scale cattle herding during the nineteenth century on the previously buffalo-dominated southern and western plains is an example of this type of adaptation within a specialized system. Swidden agriculture is certainly an example of it within a generalized one.

The extraordinarily high diversity index of the tropical forest, the kind of natural climax community which still characterizes the bulk of Outer Indonesia, has already been mentioned. Though there are probably more floral species in this region than any other of comparable size in the world (van Steenis has estimated that between twenty and thirty thousand species of flowering plants, belonging to about 2,500 families, can be found in the archipelago), continuous stands of trees or other plants are rare, and the occurrence of as many as thirty different species of trees within a hundred square yards is not at all uncommon.[10] Similarly, on about a three-acre swidden plot in the Philippines (detailed field studies are lacking for Indonesia as such) Conklin has seen as many as forty different sorts of crops growing simultaneously, and one informant drew an ideal plot containing at one time forty-eight basic kinds of plants. The people of the area, the Hanunóo of Mindoro, distinguish more than sixteen hundred different plant types (which is a finer classification than that employed by systematic botanists), including the astounding number

---

[10] van Steenis, 1935; and Dobby, 1954, p. 61. This floral diversity is paralleled by an equally great wealth of fauna: the industrious as well as famous naturalist A. R. Wallace found 200 species of beetles in a square mile of Singapore forest and brought back a total of more than 125,000 animal specimens from the general Malaysian region. Robequain, 1954, pp. 38–59. For a general ecological analysis of tropical forest plant diversity, see Richards, 1952, pp. 231–268. More popular accounts, but which include some discussion of fauna as well, are Bates, 1952, pp. 175–211; and Collins, 1959.

of four hundred thirty cultivates.[11] Conklin's vivid description of what a Hanunóo swidden in full swing looks like gives an excellent picture of the degree to which this agriculture apes the generalized diversity of the jungle which it temporarily replaces:

Hanunóo agriculture emphasizes the intercropping of many types of domesticated plants. During the late rice-growing seasons, a cross section view of a new [plot] illustrates the complexity of this type of swidden cropping (which contrasts remarkably with the type of field cropping more familiar to temperate zone farmers). At the sides and against the swidden fences there is found an association dominated by low, climbing or sprawling legumes (asparagus beans, sieva beans, hyacinth beans, string beans, and cowpeas). As one goes out into the center of the swidden, one passes through an association dominated by ripening grain crops but also including numerous maturing root crops, shrub legumes and tree crops. Pole-climbing yam vines, heart-shaped taro leaves, ground-hugging sweet potato vines, and shrublike manioc stems are the only visible signs of the large store of starch staples which is building up underground, while the grain crops fruit a meter or so above the swidden floor before giving way to the more widely spaced and less rapidly-maturing tree crops. Over the first two years a new swidden produces a steady stream of harvestable food in the form of seed grains, pulses, sturdy tubers, and underground stems, and bananas, from a meter below to more than 2 meters above the ground level. And many other vegetable, spice and nonfood crops are grown simultaneously.[12]

The second formal characteristic common to the tropical-forest and swidden-agriculture ecosystems is the ratio of the quantity

---

[11] Conklin, 1954. Other valuable field studies of swidden in Malaysia include, Freeman, 1955 (on diversity, pp. 51–54); and Geddes, 1954 (on diversity, pp. 64–65). A brief description of swidden-making in East Indonesia can be found in Goethals, 1961, pp. 25–29.

[12] Conklin, 1957, p. 147. Conklin estimates that in the first and most active year of the swidden cycle up to 150 specific crop types may be planted at one time or another.

of nutrients locked up in living forms (that is, the biotic community) to that stored in the soil (that is, the physical substratum): in both it is extremely high. Though, as with the tropical forest itself, much variation is found, tropical soils are in general extensively laterized. As precipitation in most of the humid, rain-heavy tropics greatly exceeds evaporation, there is a significant downward percolation through the soil of relatively pure, lukewarm water, a type of leaching process whose main effect is to carry away the more highly soluble silicates and bases, while leaving behind a dreary mixture of iron oxides and stable clays. Carried to an extreme, this produces ferralite, a porous, crumbly, bright-red, acidic soil which, however excellent the Indonesians find it for making bricks without straw, is of much less value from the point of view of the support of plant life. Protected to a certain extent by the shielding effects of the thick vegetation cover, most tropical soils have not developed such a serious case of what Gourou has called pedological leprosy.[13] But the great majority of them, having been exposed to these ultrastable climatic conditions over very long periods of time, are markedly leached, and thus seriously impoverished in minerals requisite to the sustenance of life.[14]

This apparent and oft-remarked paradox of a rich plant and animal life supported on a thin soil is resolved by the fact that the cycling of material and energy among the various components of a tropical forest is both so rapid and so nearly closed that only the uppermost layers of the soil are directly and significantly involved in it, and they but momentarily. The intense humidity and more or less even distribution of rainfall, the equable, moderately elevated temperatures, the small month-to-month variations in day length and amount of sunlight—all the monotonous con-

---

[13] Gourou, 1953b, p. 21.

[14] This paragraph and those immediately following are based mainly on Richards, 1952, pp. 203–26; Dobby, 1954, pp. 74–84; and Gourou, 1953b, pp. 13–24. However, much remains to be learned about soil factors in the tropics.

stancies of the tropics—are conducive to a high rate of both decomposition and regeneration of animal and vegetable material. Speedy decomposition is insured by the multiplication of bacteria, fungi and other decomposers and transformers which the humid conditions favor, as well as by the multitude of herbiverous animals and insects who are so ravenous that, as Bates remarks, virtually "every fruit and every leaf [in the tropical forest] has been eaten by something." [15] An enormous amount of dead matter is thus always accumulating on the forest floor—leaves, branches, vines, whole plants, faunal remains and wastes; but their rapid decay and the high absorptive capacity of the luxuriant vegetation means that the nutrients in this dead organic matter are reutilized almost immediately, rather than remaining stored to any great extent or for any great length of time in the soil where they are prey to the leaching process.

The role of humus in creating a topsoil storehouse of nutrient materials in colloidal form to be drawn upon gradually as needed, which looms so prominently in ecosystems at higher latitudes, is here minimized; organic materials rarely extend in significant quantity more than a few inches beneath the forest floor, because the nutrients set free by the rapid decay of dead matter are quickly taken up again by the shallow, splayed root systems of the intensely competitive plants. Thus, despite the heavy rains, loss of nutrients due to runoff in this process of transfer is very slight, so that quite marginal additions of energy from outside the system through nitrogen fixing in leguminous trees and adsorption of minerals released by rock decomposition are themselves enough to compensate for it. The climax community, once established, through still imperfectly understood processes of ecological succession, is thus virtually self-perpetuating. By maintaining most of its energy in the form of living things most of the time, the tropical-forest ecosystem is able to prevent any significant escape of energy across its boundaries and to circumvent

---

[15] Bates, 1952, p. 209.

the problem of impoverished soil conditions by feeding largely upon itself.

Swidden agriculture operates in essentially this same supernatant, plant-to-plant, direct cycling manner. The burning of the slashed plot is at base a means both of accelerating the process of decay and of directing that process in such a fashion that the nutrients it releases are channeled as fully as possible into certain selected food-producing plants. A significant proportion of the mineral energy upon which swidden cultivates, and especially the grains, draw for their growth comes from the ash remains of the fired forest, rather than from the soil as such, so that the completeness with which a plot is burnt is a crucial factor in determining its yield, a fact of which probably all swidden cultivators are aware.[16] A good burn, in turn, is dependent on the one hand upon the care and thoroughness with which the vegetation has been cut, and on the other upon the dryness of the weather during the cutting-planting period. Over the thoroughness of the cutting the cultivators have a high degree of control and, though different groups of swidden agriculturalists, as any other type of farmers, vary widely in their skills, yet their felling, slashing, trimming techniques, as well as their actual firing methods, are commonly well developed. Over the weather they have, of course, no control (though they are usually adept at estimating it), and intense ritual activity is commonly directed toward preventing rain, or at least maintaining confidence, during the anxious, all-important few weeks between cutting and sowing. At any rate, the primary function of "slash and burn" activities is not mere clearing of the land (the use of the term "clearing," with respect

---

[16] For example, among the Mandaya of eastern Mindanao, those cultivating over 1,700 feet where burning is impossible because of the absence of a dry period harvest about 10 to 15 cavans of rice per hectare, while those cultivating in lower areas where burning is possible average 30–35; Aram A. Yengoyan, personal communication. In general, however, the precise effect of firing as a fertilizing mechanism remains to be investigated experimentally, like so much else about swidden.

to swiddens is actually somewhat misleading) but rather the ✓
transfer of the rich store of nutrients locked up in the prolific
vegetation of the tropical forest to a botanical complex whose
general ecological productivity, in the sense of the total energy
flow in the system, may be substantially smaller but whose yield
to man is a great deal larger.[17]

General ecological productivity is lower because this transfer is
less efficient than that which takes place under natural conditions
of decay and regeneration. Here, a large amount of energy does
escape across the boundaries of the system. Gourou estimates that
between six and nine hundred pounds of nitrogen alone go up in
smoke in the burning of a single acre of forest; and, despite the
utmost shrewdness in judging the weather and the greatest speed
in firing and planting, much ash is inevitably washed away by
the rains before it can be utilized by the cultivates, fast growing
as they tend to be.[18] Further, as the cultivates are less woody in
substance than those indigenous to the forest, they do not form
a very appropriate material for the technique of accelerating and
channeling nutrient transfer through the deliberate production of
ash, and so the firing process is not continuously repeatable. The
result is, of course, the well-known drop in fertility on swidden
plots (rice output of south Sumatran plots is known to drop as
much as 80 percent between a first and second cropping), and the
surrender of the plot to natural regeneration.[19]

But, despite the fact that secondary forest growth is, at least
in the earlier phases of regeneration, notably less luxuriant than
primary, if the period of cultivation is not too long and the period
of fallow long enough, an equilibrated, nondeteriorating and reasonably productive farming regime (productive in the sense of

---

[17] This analysis is based on the descriptions of swidden techniques given
in Conklin, 1957, pp. 49–72; Freeman, 1955, pp. 40–48; and Hose and
MacDougal, 1912. For the distinction between ecological productivity and
yield to man, see Clarke, 1954, pp. 482–500.

[18] Gourou, 1953b, p. 26.

[19] Pelzer, 1945, p. 16.

yield to man) can be sustained, again to a significant degree irrespective of the rather impoverished soil base on which it rests.[20] The burned forest provides most of the resources for the cultivates; the decaying cultivates (nothing but the edible portions of plants is removed from the plot) and the natural processes of secondary succession, including invasions from the surrounding forest within which plots are usually broadly dispersed rather than tightly clustered, provide most of the resources for the rapidly recuperating forest. As in the undisturbed forest, "what happens" in an adapted swidden ecosystem happens predominantly in the biotic community rather than in the physical substratum.

Finally, a third systemic property in which the tropical forest and the swidden plot tend to converge is general architecture: both are "closed-cover" structures. The tropical forest has often been compared to a parasol, because of the effectiveness with which the tall, closely packed, large-crowned, evergreen trees both deflect the rain and shut out the sun so as to protect the soil against the worst effects of the leaching process, against baking, and against erosion. Photosynthesis takes place almost entirely at the very top of the forest, from a hundred to a hundred and fifty feet up, and so most of the growing things (as well as much of the faunal life) reach desperately toward this upper canopy seeking their small place in the sun, either by climbing, as the thousands of woody lianas and other vines, by finding an epiphytic perch, as the orchids and ferns, or by mere giantism, as the dominant trees and the bamboos, leaving the darkened floor relatively free of living plants.[21] In a swidden, this canopy is, of course, radically lowered, but much of its umbrella-like continuity is maintained, in part by planting cultigens not in an open field, crop-row manner, but helter-skelter in a tightly woven, dense botanical fabric, in part by planting shrub and tree crops of vari-

---

[20] Conklin, 1957, p. 152; Leach, 1954, p. 24; and Geddes, 1954, pp. 65–68.
[21] Bates, 1952, pp. 200–203.

ous sorts (coconuts, areca, jakfruit, banana, papaya, and today in more commercial areas rubber, pepper, abaca, and coffee), and in part by leaving some trees standing. In such a way, excessive exposure of the soil to rain and sun is minimized and weeding, exhausting task in any case, is brought within reasonable proportions because light penetration to the floor is kept down to a much lower level than in an open-field system.[22]

In sum, a description of swidden farming as a system in which "a natural forest is transformed into a harvestable forest" seems a rather apt one.[23] With respect to degree of generalization (diversity), to proportion of total system resources stored in living forms, and to closed-cover protection of an already weakened soil against the direct impact of rain and sun, the swidden plot is not a "field" at all in the proper sense, but a miniaturized tropical forest, composed mainly of food-producing and other useful cultivates. Yet, as is well known, though less well understood, the equilibrium of this domesticated form of forest system is a great deal more delicate than that of the natural form. Given less than ideal conditions, it is highly susceptible to breakdown into an irreversible process of ecological deterioration; that is, a pattern of change leading not to repeated forest recuperation but to a replacement of tree cover altogether by the notorious *imperata* savannah grass which has turned so much of Southeast Asia into a green desert.[24]

---

[22] For an excellent description of the concurrent employment by recent immigrant Javanese farmers of an open-field system and by indigenous farmers of a closed-field one in the Lampong area of south Sumatra, and of the essential defeat of the former by the weeding problem, see, Kampto Utomo, 1957, pp. 127–132. Some forms of partial swidden-farming—i.e., where swidden is auxiliary to other forms of cultivation—are, however, open-field systems; while integral systems—i.e., where swidden is the sole form of cultivation—commonly are not. I owe this point to Harold Conklin.

[23] Kampto Utomo, 1957, p. 129.

[24] Gourou, 1953a, p. 288, estimates that about 40 percent of the Philippines and 30 percent of Indonesia are covered with *imperata,* presumably

Swidden cultivation may turn thus maladaptive in at least three ways: by an increase in population which causes old plots to be recultivated too soon; by prodigal or inept agricultural practices which sacrifice future prospects to present convenience; and by an extension into an insufficiently humid environment in which the more deciduous forests have a much slower recovery rate and in which clearing fires are likely to burn off accidentally great stands of timber.[25] The population problem has been much discussed, though exact figures are difficult to obtain. Van Beukering has put the population ceiling for swidden in Indonesia over-all at about 50 per square kilometer, Conklin estimates that the Hanunóo area can carry 48 per square kilometer without deterioration, and Freeman calculates 20–25 as the maximum in his central Sarawak region; but it is not known to what degree the various local population densities in Outer Indonesia now exceed critical limits and are producing grassland climaxes as a result of the need for more rapid recultivation.[26] With the population of the region now increasing at 2 percent or more annually, however, the problem seems likely to become overtly pressing in the not too distant future; glib references to Outer Indonesia as "grossly underpopulated" constitute a simplistically quantitative and ecologically naive view of demography.

The fact that wasteful or inept methods may be destructive to the long-run equilibrium of swidden agriculture not only underscores the wide variation in proficiency with which different groups of shifting cultivators operate, but, even more important,

---

nearly all of it caused by man. These figures may be somewhat high, however: Pelzer, 1945, p. 19, estimates the Philippine grassland percentage at 18.

[25] A full consideration of the factors relating to the breakdown of the swidden cycle into a deflected grassland succession would need, of course, to consider topographical and edaphic variables, the role of animal husbandry, associated hunting practices, and so on. For such a micro-analysis, see Conklin, 1959.

[26] Van Beukering, 1947. Conklin, 1957, pp. 146–147. Freeman, 1955, pp. 134–135. These various figures are all somewhat differently calculated.

demonstrates that cultural, social, and psychological variables are at least as crucial as environmental ones in determining the stability of human modes of adaptation. An example of such a thriftless use of resources by swidden farmers is provided by Freeman who says that the Iban have been less shifting cultivators than *mangeurs de bois*.[27] Located in a primary forest area into which they have fairly recently expanded at the expense of indigenous tribes, the Iban are well below maximum population densities. But they nevertheless seriously overcultivate, often using a single plot three years in succession or returning to a fallowed one within five years, and thereby causing widespread deforestation. The reasons for this overcultivation are various, including an historically rooted conviction that there are always other forests to conquer, a warrior's view of natural resources as plunder to be exploited, a large village settlement pattern which makes shifting between plots a more than usually onerous task, and, perhaps, a superior indifference toward agricultural proficiency. But, again, to what degree such prodigality exists among the swidden agriculturalists of Outer Indonesia is virtually unknown.

As for the climatic factor, the most highly generalized, evergreen, closed-cover tropical forest, commonly specified as "rain forest" is chiefly characteristic of equatorial lowland areas where a marked dry season is absent; as one moves toward higher-latitude areas with a marked dry season, it shades off, more or less gradually, into a shorter, more open, less diverse, and at least partly deciduous variety of tropical forest, usually called "monsoon forest."[28] The delicacy of swidden equilibrium increases at equal pace with this transition toward a more subtropical environment because of the steadily diminishing power of the natural community rapidly to reconstitute itself after human interference. The greater ease, and uncontrollability, with which such drier wood-

[27] Freeman, 1955, pp. 135–141.

[28] Dobby, 1954, pp. 62, 65–70. Variation in tropical forest composition is also affected by altitude, soil, and local land mass configurations. For a full discussion, see Richards, 1952, pp. 315–374.

lands burn, fanned often by stronger winds than are common in the rain forest areas, only increases the danger of deterioration to grassland or scrub savannah and, in time, by erosion to an almost desert-like state. The southeast portion of the Indonesian archipelago, the Lesser Sundas, where the parching Australian monsoon blows for several months a year, has been particularly exposed to this general process of ecological decline, and in some places devastation is widespread.[29] All in all, the critical limits within which swidden cultivation is an adaptive agricultural regime in Outer Indonesia are fairly narrow.

Sawah

The micro-ecology of the flooded paddy field has yet to be written. Though extensive and detailed researches into the botanical characteristics of wet rice, its natural requirements, the techniques of its cultivation, the methods by means of which it is processed into food, and its nutritional value have been made, the fundamental dynamics of the individual terrace as an integrated ecosystem remain unclear.[30] The contrast between such a terrace —an artificial, maximally specialized, continuous-cultivation, open-field structure to a swidden plot could hardly be more extreme; yet how it operates as an organized unit is far from being understood. Knowledge remains on the one hand specialized and technical, with developed, even experimental, analyses of breeding and selection, water supply and control, manuring and weeding, and so on, and, on the other, commonsensical, resting on a vast, unexamined accumulation of proverbial, rice-roots wisdom concerning similar matters. But a coherent description of the manner in which the various ecological components of a terrace interrelate to form a functioning productive system remains noticeable by its absence. So far as I am aware, a genuinely detailed and circumstantial analysis of any actual wet-rice field (or group of fields) as a set of "living organisms and nonliving sub-

---

[29] See Ormeling, 1956.

[30] For an encyclopedic summary of such researches, see Grist, 1959.

stances interacting to produce an exchange of material between
with living and the non-living parts" does not exist in the
literature.[31]

The most striking feature of the terrace as an ecosystem, and
the one most in need of explanation, is its extraordinary stability
or durability, the degree to which it can continue to produce,
year after year, and often twice in one year, a virtually undimin-
ished yield.[32] "Rice grown under irrigation is a unique crop,"
the geographer Murphey has written,

. . . soil fertility does affect its yield, as does fertilization, but
it does not appear to exhaust the soil even over long periods without fertilization, and in many cases it may actually improve the soil.
On virgin soils a rapid decline in yield usually takes place, in the
absence of fertilization, within the first two or three years, but after
ten or twenty years the yield tends to remain stable more or less
indefinitely. This has been borne out by experiments in various
parts of tropical Asia, by increased knowledge of the processes involved, and by accumulated experience. On infertile soils and with
inadequate fertilization the field stabilizes at a very low level, as is
the case now in Ceylon and most of South Asia, but it does stabilize.
Why this should be so is not yet entirely understood.[33]

The answer to this puzzle almost certainly lies in the paramount
role played by water in the dynamics of the rice terrace. Here, the
characteristic thinness of tropical soils is circumvented through
the bringing of nutrients onto the terrace by the irrigation water
to replace those drawn from the soil; through the fixation of
nitrogen by the blue-green algae which proliferate in the warm
water; through the chemical and bacterial decomposition of organic material, including the remains of harvested crops in that
water; through the aeration of the soil by the gentle movement

---

[31] The quotation is the formal definition of an ecosystem given in Odum,
1959, p. 10.
[32] Gourou, 1953b, p. 100; and 1953a, p. 74.
[33] Murphey, 1957.

of the water in the terrace; and, no doubt, through other ecological functions performed by irrigation which are as yet unknown.[34] Thus, although, contrary to appearances, the paddy plant actually requires no more water than dry-land crops for simple transpirational purposes, "the supply and control of water . . . is the most important aspect of irrigated paddy cultivation; given an adequate and well-controlled water supply the crop will grow in a wide range of soils and in many climates. It is therefore more important than the type of soil."[35]

This primary reliance on the material which envelops the biotic community (the "medium") for nourishment rather than on the solid surface in which it is rooted (the "substratum"), makes possible the same maintenance of an effective agricultural regime on indifferent soils that the direct cycling pattern of energy exchange makes possible on swiddens.[36] Even that soil quality which is of clearest positive value for paddy growing, a heavy consistency which irrigation water will not readily percolate away, is more clearly related to the semiaquatic nature of the cultivation process than to its nutritional demands, and paddy can be effectively grown on soils which are "unbelievably poor in plant nutrients."[37] This is not to say that natural soil fertility has no effect on wet-rice yields, but merely that, as "paddy soils tend to acquire their own special properties after long use," a low natural fertility is not in itself a prohibitive factor if adequate water resources are available.[38] Like swidden, wet-rice cultivation is essentially an ingenious device for the agricultural exploitation of a habitat in which heavy reliance on soil processes is impossible

[34] In addition to the mentioned Grist (esp. pp. 28–49), Gourou, and Murphey references, useful, if unsystematic, material on the micro-ecology of irrigated rice can be found in Pelzer, 1945, pp. 47–51, and especially in Matsuo, 1955, pp. 109–12.

[35] Grist, 1959, pp. 28, 29.

[36] For the distinction between "medium" and "substratum," see Clarke, 1954, pp. 23–58, 59–89.

[37] Pendleton, 1947; quoted in Grist, 1959, p. 11.

[38] Murphey, 1957.

and where other means for converting natural energy into food
are therefore necessary. Only here we have not the imitation of a
tropical forest, but the fabrication of an aquarium.

The supply and control of water is therefore the key factor in
wet-rice growing—a seemingly self-evident proposition which
conceals some complexities because the regulation of water in a
terrace is a matter of some delicacy. Excessive flooding is often as
great a threat as insufficient inundation; drainage is frequently
a more intractable problem than irrigation. Not merely the gross
quantity of water, but its quality, in terms of the fertilizing sub-
stances it contains (and thus the source from which it comes) is
a crucial variable in determining productivity. Timing is also
important: paddy should be planted in a well-soaked field with
little standing water and then the depth of the water increased
gradually up to six to twelve inches as the plant grows and
flowers, after which it should be gradually drawn off until at
harvest the field is dry. Further, the water should not be allowed
to stagnate but, as much as possible, kept gently flowing, and
periodic drainings are generally advisable for purposes of weeding and fertilizing.[39] Although with traditional (and in some
landscapes, even modern) methods of water control the degree to
which these various optimal conditions can be met is limited,
even at its simplest, least productive, and most primitive this
form of cultivation tends to be technically intricate.

And this is true not only for the terrace itself, but for the system of auxiliary water works within which it is set. We need not
accept Karl Wittfogel's theories about "hydraulic societies" and
"oriental despotisms" to agree that while the mobility of water
makes it "the natural variable *par excellence*" in those landscapes
where its manipulation is agriculturally profitable, its bulkiness
makes such manipulation difficult, and manageable only with
significant inputs of "preparatory" labor and at least a certain

---

[39] Grist, 1959, pp. 28–32. One of the primary functions, aside from nutrition, of irrigation water is, in fact, the inhibition of weed growth.

amount of engineering skill.[40] The construction and maintenance
of even the simplest water-control system, as in rainfall farms,
requires such ancillary efforts: ditches must be dug and kept
clean, sluices constructed and repaired, terraces leveled and dyked;
and in more developed true irrigation systems dams, reservoirs,
aqueducts, tunnels, wells and the like become necessary. Even
such larger works can be built up slowly, piece by piece, over extended periods and kept in repair by continuous, routine care.
But, small or large, waterworks represent a level and kind of
investment in "capital equipment" foreign not only to shifting
cultivation but to virtually all unirrigated forms of premodern
agriculture.

This complex of systemic characteristics—settled stability, "medium" rather than "substratum" nutrition, technical complexity
and significant overhead labor investment—produce in turn what
is perhaps the sociologically most critical feature of wet-rice agriculture: its marked tendency (and ability) to respond to a rising
population through intensification; that is, through absorbing
increased numbers of cultivators on a unit of cultivated land. Such
a course is largely precluded to swidden farmers, at least under
traditional conditions, because of the precarious equilibrium of the
shifting regime. If their population increases they must, before
long, spread out more widely over the countryside in order to
bring more land into cultivation; otherwise the deterioration to
savannah process which results from too rapid recultivation will
set in and their position will become even more untenable. To
some extent, such horizontal expansion is, of course, possible for
traditional wet-rice agriculturalists as well, and has in fact
(though more slowly and hesitantly than is sometimes imagined)
occurred. But the pattern of ecological pressures here increasingly
encourages the opposite practice: working old plots harder rather
than establishing new ones.

The reasons for this introversive tendency follow directly from

---

[40] Wittfogel, 1957, p. 15.

the listed systemic characteristics. The stability of the rice terrace as an ecosystem makes the tendency possible in the first place. Because even the most intense population pressure does not lead to a breakdown of the system on the physical side (though it may lead to extreme impoverishment on the human side), such pressure can reach a height limited only by the capacity of those who exploit it to subsist on steadily diminishing per capita returns for their labor. Where swidden "overpopulation" results in a deterioration of the habitat, in a wet-rice regime it results in the support of an ever-increasing number of people within an undamaged habitat. Restricted areas of Java today—for example, Adiwerna, an alluvial region in the north-central part of the island—reach extraordinary rural population densities of nearly 2,000 persons per square kilometer without any significant decline in per-hectare rice production. Nor does there seem to be any region on the island in which wet-rice growing was employed effectively in the past but cannot now be so employed due to human over-driving of the landscape. Given maintenance of irrigation facilities, a reasonable level of farming technique, and no autogenous changes in the physical setting, the *sawah* (as the Javanese call the rice terrace) seems virtually indestructible.

Second, the "medium-focused" quality of the regime limits it fairly sharply to those areas in which topography, water resources, and soluble nutrients combine to make the complex ecological integration of sawah farming (whatever that may turn out in detail to be) possible. All agricultural regimes are, of course, limited by the environmental conditions upon which they rely. But wet-rice cultivation, particularly under premodern technological conditions, is perhaps even more limited than most and, within Indonesia, certainly more than swidden, which can be carried out over the greater part of the archipelago, including, as it once was, most of those parts now pre-empted by sawah. Swidden can be pursued on rugged hillsides, in wet lowland forests, and in relatively dry monsoon country where, at least without the assistance of modern methods of water control, con-

servation, and regulation, sawah cannot. Exact data are difficult
to obtain but the great extension of irrigated rice-farming in
Indonesia and the rest of Southeast Asia during the last hundred
years or so as a result of the application of Western technology
ought not to obscure the fact that before the middle of the nineteenth century such farming was restricted to a few, particularly
favorable areas. In 1833, when Java was just on the eve of her
most disastrous period of social change, the island, which today
has about three and a half million hectares of sawah had only
slightly more than a third that much.[41]

Yet there is another introversive implication of the technical
complexity aspect of traditional wet-rice cultivation. Because productivity is so dependent on the quality of water regulation,
labor applied to the improvement of such regulation can often
have a greater marginal productivity than that same labor applied
to constructing new, but less adequately managed, terraces and
new works to support them. Under premodern conditions,
gradual perfection of irrigation techniques is perhaps the major
way to raise productivity not only per hectare but per man. To
develop further water works already in being is often more profitable than to construct new ones at the established technical level;
and, in fact, the ingenious traditional water-control systems of
Java and Bali can only have been created during a long period
of persistent trial-and-error refinement of established systems.
Once created, an irrigation system has a momentum of its own,
which continues, and even increases, to the point where the
limits of traditional skills and resources are reached. And, as the
gap between the first rainfall, stream-bank, or swamp-plot sawah
and those limits is usually great, economic progress through step-
by-step technological advance within a specifically focused system
can be an extended process, as shown in the following description
of a Ceylonese system:

----

[41] The contemporary figure is from Statistical Pocketbook of Indonesia,
1957, p. 46; the 1833 figure (1,270,000 ha.) from van Klaveren, 1955, p. 23.

. . . the Kalāwewa canal system—now has a giant tank at its head which leads into a fifty-five mile long watercourse, which in turn feeds into three large tanks which provide water for the ancient capital of Anuradhapura. It all looks like a colossal and highly organized piece of bureaucratic planning, the work of one of Witt-fogel's idealised Oriental Despots. But if so, the planning must have been done by a kind of Durkheimian group mind! The original Tissawewa tank at the bottom end of the system was first constructed about 300 B.C. The Kalāwewa tank at the top end of the system was first constructed about 800 years later and elaborations and modifica-tions went on for at least another 600 years.[42]

However, as mentioned, it is not only with respect to ancillary waterworks that wet-rice agriculture tends toward technical com-plexity, but on a more microscopic level with respect to the indi-vidual terrace itself. In addition to improving the general irriga-tion system within which a terrace is set, the output of most terraces can be almost indefinitely increased by more careful, fine-comb cultivation techniques; it seems almost always possible somehow to squeeze just a little more out of even a mediocre sawah by working it just a little bit harder. Seeds can be sown in nurseries and then transplanted instead of broadcast; they can even be pregerminated in the house. Yield can be increased by planting shoots in exactly spaced rows, more frequent and com-plete weeding, periodic draining of the terrace during the grow-ing season for purposes of aeration, more thorough ploughing, raking, and leveling of the muddy soil before planting, placing selected organic debris on the plot, and so on; harvesting tech-niques can be similarly perfected both to reap the fullest percent-age of the yield and leave the greatest amount of the harvested crop on the field to refertilize it, such as the technique of using the razor-like hand blade found over most of inner Indonesia; double cropping and, in some favorable areas, perhaps triple cropping, can be instituted. The capacity of most terraces to re-spond to loving care is amazing. As we shall see, a whole series

---

[42] Leach, 1959.

of such labor-absorbing improvements in cultivation methods have played a central role in permitting the Javanese rural economy to soak up the bulk of the island's exploding population.

Finally, independently of the advantages of technical perfection, the mere quantity of preparatory (and thus not immediately productive) labor in creating new works and bringing them up to the level of existing ones tends to discourage a rapid expansion of terraced areas in favor of fragmentation and more intensive working of existing ones. In developed systems, this is apparent; a people who have spent 1,400 years in building an irrigation system are not likely to leave it readily for pioneering activities, even if the established system becomes overcrowded. They have too much tied up in it, and at most they will gradually create a few new terraces on the periphery of the already well-irrigated area, where water resources and terrain permit. But this reluctance to initiate new terrace construction because of the heavy "overhead" labor investment is characteristic even of areas where irrigation is still undeveloped, because of the inability or the unwillingness of peasants to divert resources from present production. In contemporary Laos, for example,

Most villagers are only semi-permanent and forest land is still available. The irrigated rice fields have become fragmented because their yields are more reliable than those of the [swidden]. The creation of new [sawahs] is not easily done, for it involves the extension of irrigation ditches and major investment of labor. This labor must be hired or supplied by the family itself, and implies existing fluid capital or a large extended family containing a number of able-bodied workers. Neither of these situations commonly occurs among Lao peasants, and therefore the progressive division of existing [wet rice] land and cultivation of [swidden] which requires less initial labor.[43]

Therefore, the characteristics of swidden and sawah as ecosystems are clear and critical: On the one hand a multicrop, highly diverse regime, a cycling of nutrients between living forms, a

---

[43] Halpern, 1961.

closed-cover architecture, and a delicate equilibrium; on the other, on open-field, monocrop, highly specialized regime, a heavy dependency on water-born minerals for nutrition, a reliance on man-made waterworks, and a stable equilibrium. Though these are not the only two traditional agricultural systems in Indonesia, they are by far the most important and have set the framework within which the general agricultural economy of the country has developed. In their contrasting responses to forces making for an increase in population—the dispersive, inelastic quality of the one and the concentrative, inflatable quality of the other—lies much of the explanation for the uneven distribution of population in Indonesia and the ineluctable social and cultural quandaries which followed from it.

PART II

The Crystallization of the Pattern

THE CLASSICAL PERIOD

How has this uneven distribution come about? Why is virtually all of the nation's swidden found today in Outer Indonesia and more than three-quarters of her sawah concentrated in the inner core of northwest, east, and central Java, south Bali and west Lombok? [1] What factors shaped this peculiar pattern of land use in the first place and what factors acted upon that pattern from outside, complicating it, solidifying it and forcing it into its present untractable, overdriven state? What is the ecological history of the ossification of the Indonesian agrarian economy?

The flourishing of wet-rice agriculture on Java (since before the Christian era) has generally been explained by a happy combination of, as the Dutch geologist Mohr put it, the four elements of the ancient world: fire, water, earth and air. [2] The "fire" is provided by the intense volcanic activity of the more than thirty working cones, which run lengthwise down the middle of the island and supply the plant nutrients which the otherwise thin

---

[1] The swidden distribution is apparent from mere inspection; the sawah proportions are calculated from Statistical Pocketbook of Indonesia, 1957, pp. 46–49.

[2] Mohr, 1946.

soils lack. The "water" comes from the short, quick-running, silt-laden rivers cradled between these cones which, draining one way or another from the central range, carry the minerals it produces southward toward the Indian Ocean or northward toward the Java Sea. The "earth" is represented by the well-drained, gradually sloping, enclosed-plain relief formed by the basins of these intermountain rivers which creates a series of well-defined natural amphitheatres eminently suited to traditional gravity-feed irrigation techniques. And the "air" is an outcome of the moderately humid climate intermediate between the continual rainfall of equatorial Sumatra, Borneo, Celebes, and the Moluccas and the sharp biseasonality of the monsoonal Lesser Sundas, a tropical compromise which avoids both the swamping and intense leaching of the islands to the north and west and the excessive drying and wind erosion of those to the south and east. Traditionally at least, Java (with Bali and Lombok) has formed not one huge continous, dead flat river-plain rice belt, as, say, central China or northwest India, but a number of separate, gently rounded pockets of intense cultivation—a set of small-scale, richly alluvial galleries hemmed in by volcanic mountains or unirrigable limestone hills.[3] (See Map 2.)

With the partial exception of the Agam and Toba regions of central and north Sumatra, such nicely appropriate landscapes for wet-rice cultivation do not exist in the Outer Islands. Sumatra is sharply divided into precipitous western highlands and spongy eastern swamp; Borneo is poorly drained on the coast, covered with broken hills in the interior; Celebes is mountain-crowded, with few sizable lowland areas; the Moluccas are fragmented and very wet, the Lesser Sundas fragmented and, in the summer monsoon, very dry. Sumatra has some active volcanoes, but most of them throw off acidic rather than basic ejecta and so impover-

---

[3] The best synthetic works on Indonesian geography are Dobby, 1954, and Robequain, 1954. For Java alone, Lekkerkerker, 1938, I, 13–90; and Veth, 1912, Vol. III.

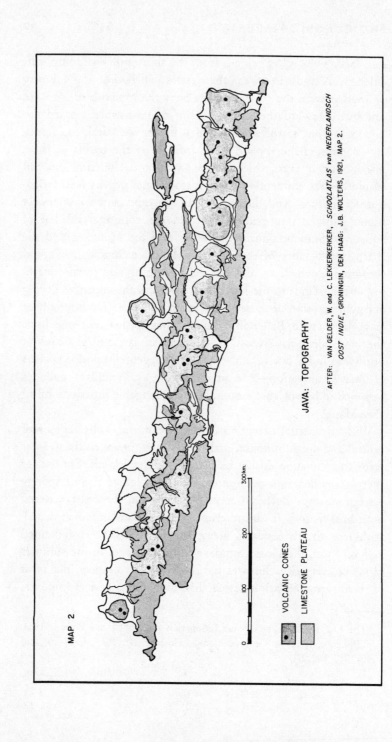

MAP 2

JAVA: TOPOGRAPHY

AFTER: VAN GELDER, W. and C. LEKKERKERKER, *SCHOOLATLAS van NEDERLANDSCH OOST INDIE*, GRONINGIN, DEN HAAG: J.B. WOLTERS, 1921, MAP 2.

VOLCANIC CONES

LIMESTONE PLATEAU

0   100   200   300 km.

ish rather than enrich the land around them, while Borneo, Celebes (except for its northern tip), and the Moluccas (except for western Halmahera) have no active volcanoes.

These climatological, topographical, and geological conditions do not necessarily preclude wet-rice agriculture from these regions, even with only traditional techniques. Its narrowly specialized ecological requirements can be met in a great diversity of settings and through a wide variety of means: the Tonkin Delta has no volcanoes, Malaya is as wet as Sumatra or Borneo, and northern Luzon is mountainous. But there is no doubt that in imposing a complex of adverse conditions for sawah to overcome, these geographical realities have played a great part in discouraging its establishment in the Outer Islands in favor of the more broadly applicable swidden; nor that the natural advantages of Java's small, volcano-rimmed river passages strongly encouraged its implantation there.

The first really effective integral wet-rice regimes seem to have been set up in a number of such passages in the central and eastern parts of the island. Perhaps the earliest of them appeared in and around the majestic volcanic quadrangle formed by Mounts Sumbing, Sindoro, Merbabu, and Merapi in the narrowed middle neck of the island; that is, along the southward-running Progo river in what is today the Magelang area, the upper Solo (Dengkeng) river southwest of Surakarta, the Seraju valley above present-day Banjumas, and the Lukolo and Bagawanta piedmonts around Kebumen and Purwaredja (Kedu)—a region which after the eighth century blossomed into the so-called Mataram empire. Somewhat later, evidently, developed sawah appeared along the upper and middle Brantas river around Malang and Kediri, and, less extensively, in the Ponorogo area south of Madiun and the Lumadjang area east of Malang—a region which flourished, as did south Bali with which it had important connections, after the tenth century. (See Map 3.)[4] Further archeological work may

---

[4] The best, and virtually the only, systematic attempt to relate the admittedly nonagricultural, archeological remains of the Hindu period to

render parts of this picture inaccurate in detail; but the general pattern is clear. The earliest manifestations of developed wet-rice agriculture almost certainly occurred either in the enclosed inner reaches of the somewhat larger northward-flowing rivers or in the upper basins of the generally shorter southward ones.

The expansion of sawah cultivation beyond the boundaries of these especially favorable regions (which are sometimes collectively referred to as Kedjawén, or "Java Proper") in precolonial times, and in fact up to the middle of the past century, was gradual, tentative, fluctuating, and only partial.

Essentially, such extension could take place in three directions (see Map 4): northward toward the deltaic Java Sea coast (the south coast is about 80 percent calcareous), westward toward the Sunda highlands, and eastward toward the drier regions of the so-called "East Hook" (Pasuruan, Probolinggo, Besuki, and Banjuwangi). Each of these areas posed different technical problems in irrigation, all of them difficult to solve with traditional methods. In the littoral, which the Javanese call the *pasisir* ("strand," "coastline"), the problem was mainly one of perfecting water control, that is, improving flood protection and drainage, because of the great size of the river discharges (several times those of comparable temperate latitude rivers) and their extraordinary month-to-month, sometimes even day-to-day, variability.[5] Fertil-

---

the Javanese landscape in order to draw some socioeconomic conclusions is Schrieke, 1957, pp. 102–104, 288–301. Some interesting speculations about Javanese agricultural history, based on distributional evidence can be found in Terra, 1958. For a general review of Indonesian post-neolithic archeology, see Bernet Kempers, 1959.

[5] The Tjimanuk at Indramaju has a normal highest discharge rate of 25,000 cubic feet per second, a normal low of 600, with occasional floods of 34,000; the Pemali at Brebes runs between 25,000 and 250, with crests at 34,000; the Solo ranges between 70,000 and 810, with 90,000 floods (!). On the general problem of the enormous river fluctuations in Java in terms of its implications for irrigation, see van der Meulen, 1949–50. Also, Dobby, 1954, pp. 47–60, 225.

MAP 3

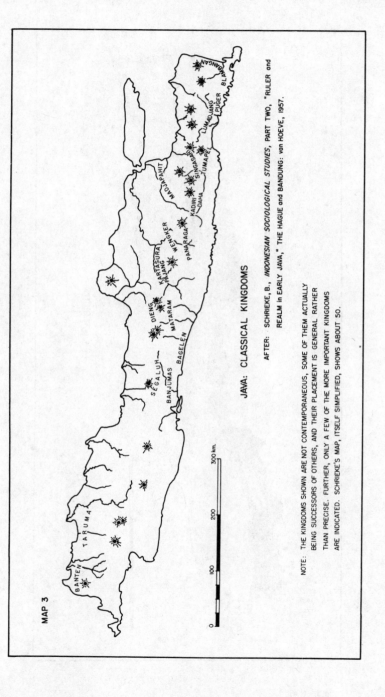

JAVA: CLASSICAL KINGDOMS

AFTER: SCHRIEKE, B., *INDONESIAN SOCIOLOGICAL STUDIES*, PART TWO, "RULER and
REALM in EARLY JAVA," THE HAGUE and BANDUNG: van HOEVE, 1957.

NOTE: THE KINGDOMS SHOWN ARE NOT CONTEMPORANEOUS, SOME OF THEM ACTUALLY
BEING SUCCESSORS OF OTHERS, AND THEIR PLACEMENT IS GENERAL RATHER
THAN PRECISE. FURTHER, ONLY A FEW OF THE MORE IMPORTANT KINGDOMS
ARE INDICATED. SCHRIEKE'S MAP, ITSELF SIMPLIFIED, SHOWS ABOUT 50.

BANTEN
TARUMA
SEGALUH
BANJUMAS
BAGELEN
DIENG
MATARAM
KARTASURA
PAJANG
WENGKER
PANARAGA
MODJAPAHIT
KADIRI
DAHA
SINGASARI
TUMAPEL
LUMADJANG
PUGER
BLAMBANGAN

0   100   200   300 km.

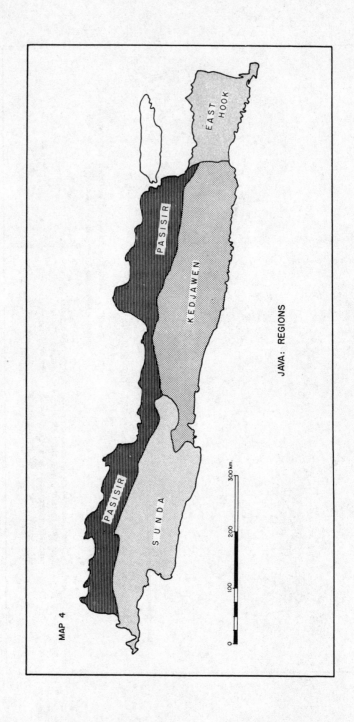

MAP 4

PASISIR

PASISIR

SUNDA

KEDJAWEN

EAST
HOOK

JAVA: REGIONS

0   100   200   300 km.

ity, particularly in the plain between Tjirebon and Djapara and in the Solo and Brantas deltas, was high, because of heavy alluvial deposits of volcanic and other sediment, and water was all too plentiful; but the great susceptibility to swamping because of near sea-level relief and the uncontrollable river spates made it a treacherous landscape for sawah during the classical period and well into the colonial. The Sunda highlands had sufficient water and good drainage, but also lower fertility because of increased leaching and less felicitous topographical arrangements. And the east was pressed for water.

As Mohr has noted, there are three independent aspects or functions of irrigation which are often insufficiently distinguished: the provision of moisture to otherwise arid soils—the "watering" function; the regulation of a quantitatively sufficient but uncertain and refractory water supply so as to avoid flooding—the "control" function; and the enrichment of the field through the transportation of nutrients to it—the "fertilization" function.[6] Which aspect is critical depends upon which factor in the environment is limiting with respect to the regime concerned.[7] Thus, where the availability of water is the limiting element, as in the East Hook, the first aspect is problematic. Where rainfall or river volume are irregular and unpredictable and drainage is difficult, as along the north coast, the second aspect looms as central. And where plant food is in short supply, as in Sunda, the third aspect is cardinal.[8] In spreading out from its enclosed-plain nurseries in the Kedjawén, wet-rice agriculture had not just a single, uniform problem to solve, but, even in the broadest terms, three technically diverse ones. That it proceeded slowly is understandable.

---

[6] Mohr, 1946.
[7] On the concept of the limiting factor, "the factor that first stops the growth and spread of [an organic system]," see Clarke, 1954, p. 20.
[8] Mohr, 1946.

Westward, in Sunda, it never got particularly far. The first reports of sawah penetration into the highlands come as late as 1750 from the Sumedang and Tasikmalaya areas, small piedmont river valleys near the eastern edge of the highlands, and a half-century later from the somewhat larger and higher plateau basins of Bandung and Bogor to the west.[9] But even today intensive sawah cultivation is limited to these areas plus a few other favored pockets (Tjiandjur, Sukabumi, the Garut plateau, and others). Eastward, in the Hook, penetration began earlier, around 1200. But as river water was in short supply the sawahs were dependent upon what rainfall could be trapped within their dikes and so remained limited in extent, reliability, and productivity until Dutch plantation interests led to the construction of modern reservoirs around the beginning of this century.[10] Even then, though its productivity has improved, the scope of wet-rice cultivation has not grown especially great, occupying, as in Sunda, around 15 percent of the total land area, as against about 25 percent in the Kedjawén and 35 percent along the Pasisir.[11]

The situation in the Pasisir, the north coast, is more complicated but the result is about the same. Some wet-rice farming has, in all probability, been carried on there at various points, particularly around the hundreds of minute estuaries which perforate the otherwise unvaried shore, since the earliest days. Some authorities

---

[9] Terra, 1958.

[10] Terra, 1958, and van der Meulen, 1949–50.

[11] These figures calculated from Landbouwatlas van Java en Madoera, 1926, I, pp. 1*–55*, Sunda being considered to consist of Banten, Bogor, and Priangan; Pasisir of Djakarta, Tjirebon, Pekalongan, Semarang, Djapara-Rembang, Bodjonegoro, and Surabaja; Kedjawén of Banjumas, Kedu, Jogia, Solo, Madiun, and Kediri; and the East Hook of Malang and Besuki. These are 1920 figures, but although the total cultivated area has probably grown since that time, the inter-regional proportions have not radically changed. The comparable figure for south Bali (Tabanan, Badung, Gianjar, Bangli, and Klungkung) is 26 percent; see Raka, 1955, p. 28.

have even speculated that the first footholds of sawah cultivation on the island were secured in these small river mouths between the larger, unmanageable deltas.[12] Yet, in a more general sense, the coast always seems to have been a chancy and difficult environment for traditional sawah; it could exist there, but not thrive.

Because of the severe water-control problems (and perhaps for other reasons as well—the malarial unhealthfulness of the region, its vulnerability to military attack), the north never became the center of a developed agrarian culture comparable to that of the interior. After the inland civilization flourished, it moved hesitantly toward the coast, drawn evidently by the attractions of a burgeoning Java Sea trade economy; but it never really arrived. Though the inland civilization periodically controlled the harbor towns, it was never able to secure that control by establishing the sort of developed wet-rice economy which supported its power in the interior. In the few places where some success was achieved— as at Demak and, perhaps, at Tjirebon—it was partial, aberrant, and short-lived, a mere interlude of limited agricultural development based on, if not shifting sands, then surging waters, and the overwhelming supremacy of the interior in agrarian matters soon reasserted itself. When the Dutch arrived in the seventeenth century, more than three hundred years after this wistful drift toward the coast began, and when the major ports, now Islamic and only partly Javanese, were in full commercial swing, the area in general was nonetheless sparsely settled, and the agricultural center of gravity was still well inland, in the regions of the old Mataram.[13] "The aforesaid [sultan of] Mataram has his residence about five or six days journey from Djapara, in the interior," wrote Jan Pieterszoon Coen, Holland's East India empire builder, "where there are diverse large, populous towns, and

[12] Terra, 1958.
[13] Schrieke, 1955, p. 265.

the land is excellently abundant in rice and other victuals; all the rice that is carried along the whole coast of Java to Molucca, Johore, Ambon and Banda is usually loaded from here." [14]

[14] Schrieke, 1955, p. 267. The Mataram here is of course the Islamic one of the seventeenth century, not the Hindu one of the eighth; but the region in which it was situated is more or less the same.

4. THE COLONIAL PERIOD: FOUNDATIONS

The Company

At the inception of the colonial period, then, the over-all ecological pattern was fairly well set: on Java, a wet-rice agrarian heartland shading off into less developed regions to the west, east, and north; in the Outer Islands, an immense tropical forest worked only here and there by small tribes of swidden farmers. The first object of interest of the Dutch, as of the Portuguese who immediately preceded them, were the Moluccas, the fabled spice islands; but their attention soon turned toward Java, and it is upon it that they mainly superimposed their colonial economy, turning back again to the Outer Islands only toward the end of the past century.

"Superimposed" is the proper word, because what the Dutch were essentially concerned to do, from 1619 to 1942, was to pry agricultural products out of the archipelago, and particularly out of Java, which were saleable on world markets without changing fundamentally the structure of the indigenous economy. The Netherlands was never able, particularly after William I's attempt to re-absorb Belgium failed, to develop a manufacture export economy even remotely comparable to that of Britain, and so the interest of the Dutch in Indonesia remained overwhelmingly mercantilist to the end. The stimulation in Indonesia of extensive markets for industrial goods, it was feared, would lead only to increased British (or, later, Japanese) influence; the essential economic task was to maintain a decent differential between the import and re-export prices of East Indian agricultural products— a task which implied the developing of Dutch commercial institutions and the discouraging of Indonesian ones.[1] Amid the apparent fluctuations of policy, the colonial period consists, from the economic point of view, of one long attempt to bring Indonesia's crops into the modern world, but not her people.

The means for accomplishing this effort to keep the natives native and yet get them to produce for world markets was the formation of a chronically, and in fact intrinsically, unbalanced economic structure sometimes referred to as "dual." [2] In the export sector, there was administrative capitalism: a system in which the holders of capital, the Dutch, regulated selling prices and wages,

---

[1] The one exception, and that but a partial one, to this generalization was the export of textiles to Indonesia from Twente after 1870. See van Klaaveren, 1955, pp. 133–136, 138, 164, 192. Of course, East India products did not have to be carried to Holland before being "re-exported," but could be taken directly to foreign ports (from 1928–1939 about 90 percent of them were—see Boeke, 1947, p. 105).

[2] Boeke, 1953. The fact that Boeke's theoretical explanations for dualism were largely unsound, his pessimistic assessment of its policy implications arbitrary, and his views concerning Indonesian (or "Eastern") "mentality" fanciful (see Higgins, 1956) ought not to obscure the fact, as it sometimes has, that, although mal-integration of labor-intensive and capital-intensive sectors is a general phenomenon, in the Netherlands East Indies economy this mal-integration was present to an extraordinarily high degree; and that Boeke recognized this fact, even if he did not understand the reasons for it, as early as his 1910 Leiden dissertation, long before the modern concern with "factor proportions," "multi-sector models" and "discontinuous investment functions" made it seem like an analytical commonplace. For the half-century debate between Boeke and his critics in Holland, the bulk of it rather beside the point, see Indonesian Economics, 1961.

controlled output, and even dictated the processes of production. In the domestic sector there was family-unit agriculture, a little home industry, and some petty internal trade. As the first expanded, stimulated by rising world commodity prices, the second contracted; land and labor were taken out of rice and other village staples and put into sugar, indigo, coffee, tobacco, and other commercial crops. As the first contracted, responding to collapsing international markets, the second expanded; and a steadily growing peasant population attempted to compensate for a lost money income, to which it had become increasingly accustomed, by intensified production of subsistence crops.

With perfect flexibility, such a system might seem less disruptive of traditional society than the more direct methods of other colonial governments—and some disinterested observers, for example Furnivall, have held it to be so.[3] But it did not even approach perfect flexibility: the monopolistic nature of economic organization on the dynamic European side, fused with the political administration of the colony (at first explicitly, later merely *de facto*), made stickiness inevitable. Not only did a booming export market tend to compress the subsistence sector beyond realistic limits, but the ability of Dutch economic leadership to respond effectively to rapidly changing market conditions and developing technology was constrained by the weight of financial conservatism and by the political and economic vicissitudes of the home country. The tendency was to maintain outmoded policies to the point where the peasant subsidization of the export sector through low rents and wages became so altogether oppressive that articulate Dutch moralists would arise to stir Holland's Calvinist conscience with the specter of "declining native welfare." Then, after a few decades of ineffectual reformist experimentation and increasing penetration of East Indies trade by rival powers, a new program would be produced which would promise, finally, to make it possible to draw a commercial

---

[3] Furnivall, 1948.

profit from Indonesia while leaving its inhabitants unharmed—
even, in any fundamental sense, untouched.

However, it was always more or less the same program in, for
the times, modern guise. East Indian colonial history was marked
by a series of politico-economic devices (the East India Company,
the Culture System, the Corporate Plantation System) by means
of which the European "merchant capitalism" side of the dual
economy was to be more efficiently organized for the production
and marketing of export crops, and the Indonesian "peasant
household" side was to be better protected against the disruptive
effects of this large-scale commercial agriculture. Driven on by
ever-increasing capital requirements, the Dutch moved from the
institutional contrivances of adventurous capitalism in the eighteenth century, to those of state capitalism in the nineteenth, and
to those of bureaucratic capitalism in the twentieth. But, as each
contrivance or device, building upon the ruins of its predecessor,
entailed a yet deeper penetration of the rural economy by Western
enterprise, it actually made it more difficult to isolate native life
from the economic forces with which such enterprise deals.
Modern Indonesia was created both because of Dutch policies and
in spite of them.

The Dutch East India Company, the first of these devices, was
formed in 1602 as a state-chartered far-eastern trade syndicate
with considerable autonomy ("a state within a state") in order to
counter the active competition of both Asiatic merchants and
other European powers trafficking around the archipelago.[4] At
first, the Company was only interested in commerce—in securing
by hook or by crook whatever products might be carried from
places where they abounded to those where they were scarce.
Aside from its depredations in the Moluccas, something of a
landmark in the history of mercantile brutality, its initial impact

---

[4] The Company's activities were of course not confined to the Indonesian
area. The best book on its commercial aspects is Glamann, 1958. For its
social impact in Indonesia, see Gonggrijp, 1957, pp. 181–321.

was mainly concentrated, therefore, in the passage ports of the Java Sea, and particularly in those dotting Java's north coast. But, in an age of rampant mercantilism, the catchword is not "business is business" but "trade follows the flag": commercial development implies political expansion. The Company—"tired of the forcing up of tolls and market taxes and the constant giving of presents to rulers and lords"—soon turned toward gaining a more comprehensive control over sources of supply.[5] By 1684 it dominated all of Sunda; by 1743 the entire Pasisir and most of the East Hook; by 1755 the Kedjawén. (The Moluccas had been in hand since about 1660.) What began as a trade combine ended, not without struggle, as a sovereign.

The colonial economy this marriage of the economic and the political produced was one in which, as Gonggrijp has well said, sharp distinctions among free, pre-emptive, quota and monopoly commerce, all conducted under the shadow of the cannon, are not too usefully made.[6] The Company, adjusting to local conditions, functioned in diverse ways in different parts of its realm. But its activities everywhere worked toward the same end: the reduction of indigenous chiefs to dependents and the substitution of tribute for trade.

In the Moluccas, the luckless source of cloves and nutmeg, the Dutch imposed restriction of cultivation, collective punishment (for "smuggling"), and forced labor exercised through the agency of humbled native rulers. In the pepper areas, Bantam-Lampong and, to a lesser extent, central Sumatra, treaties with harbor sultans established quotas and fixed prices. In the Priangan highlands coffee gardens were introduced with traditional aristocrats acting as the Company's labor contractors. Immediately around Batavia and the adjoining northwest coast, there were nearly a hundred private sugar estates, leased from local lordlings now converted to Company employees, the proprietors of which (almost all Chinese) consequently exercised seignorial rights over

---

[5] Schrieke, 1955, p. 62.
[6] Gonggrijp, 1957, p. 40.

the villagers who chanced to live on them. And in still but half-subdued central Java were simple levies in rice, timber, cotton thread, beans, and cash. By the time the Company dissolved in bankruptcy on the last day of the eighteenth century, it had laid down the general lines which the Netherlands East Indies economy followed to the end.

But only the general lines. The Company established the Dutch presence in the archipelago, introduced a few new cultivations (mainly coffee; pepper, spices, and sugar preceded it) and devised, in rough form, most of the techniques for skimming cash crops off the surface of an immobilized subsistence economy which later became so useful when Java filled up with peasants. But its impact on the Indonesian ecological pattern as a whole was marginal and unsystematic. It capitalized on it where it could, demanding deliveries, fixing prices, restricting trade but, with a few exceptions, such as the decimation of Banda, it did not attempt to act upon it directly. Its work was pioneering; consolidation of the approach it established came after it with the next major scheme to make Indonesia pay—the Culture System.

The Culture System

To speak of the Culture System[7] as a stage of Indonesian economic history, however, is to speak synecdochically. That stroke of fiscal genius by Governor General van den Bosch which we conceive to be the System proper—the remission of the peas-

---

[7] Cultuurstelsel. Properly, this term ought to be Englished as "Cultivation System," but the "Culture System" mistranslation is so embedded in the literature that it seems less confusing to continue to employ it. A great deal has been written about the Culture System (see, for example, the bibliography in Reinsma, 1955, pp. 183–189), but much of it has been marred by focus on its short-run impact at the expense of its long-run effects, and in particular by an anxious concern with its immediate moral justifiability (or lack thereof) rather than its importance as part of the crystallizing Colonial pattern in Indonesia and the role it consequently played in the formation of modern Indonesian society. The most important exception to this stricture is the discussion by Burger, 1939, pp. 117–160.

ant's land taxes in favor of his undertaking to cultivate government-owned export crops on one-fifth of his fields or, alternatively, to work sixty-six days of his year on government-owned estates or other projects—was only part of a much larger complex of politico-economic policies and institutions. Alongside the mammoth state plantation which this system tended to make of Java, there was a whole series of adjuncts, related systems, and independent growths, so that the picture of the island from 1830 to 1870 is a much more differentiated and much less static one than that which has so often been drawn for us. And yet, in spite of this, in spite of the fact that it never encompassed more than about 6 percent of Java's cultivated land or about a quarter of her people in any one year, and although it was only fully applied for perhaps two decades, van den Bosch's miraculous invention ("less taxes but more Government revenue!" as de Graff exclaims in mock wonder) does define the period.[8] And the period defines the age: ecologically, at any rate, it was the most decisive of the Dutch era, the classic stage of colonial history, as the Company was the formative.

The System, in this larger sense, was decisive in at least three ways. By its intense concentration on Java it gave a final form to the extreme contrast between Inner Indonesia and Outer which thenceforth merely deepened. It stabilized and accentuated the dual economy pattern of a capital-intensive Western sector and a labor-intensive Eastern one by rapidly developing the first and rigorously stereotyping the second, a gulf which also subsequently merely widened as Dutch investment grew. And, most important of all, it prevented the effects on Javanese peasantry and gentry alike of an enormously deeper Western penetration into their life from leading to autochthonous agricultural modernization at the point it could most easily have occurred. Such charges, more serious than those usually leveled against the System by its enemies (that it was brutal, corrupt or uneconomic), demand, clearly, both explication and substantiation.

---

[8] de Graff, 1949, p. 407.

The impact of the Culture System upon indigenous Javanese agriculture was, of course, mainly exercised through the agency of the cultivations it imposed as substitutes for money taxes.[9] Under its aegis, virtually every crop which at the time might conceivably be grown with profit was attempted: indigo, sugar, coffee, tea, tobacco, pepper, cinchona, cinnamon, cotton, silk, cocheneale; in Tjirebon, the government even tried for a while, with disastrous results, to turn rice into a levy by demanding it in lieu of taxes also and contracting with private millers to mill it.[10] Almost all these experiments, save coffee and, most spectacularly, sugar, eventually failed; but the effect of this tinkering with the established ecosystems was nonetheless profound. Today almost none of these crops are of central importance in Indonesian exports, which—minerals aside—are dominated by rubber and copra. But they established the matrix within which the present farming system, steadily replacing profitless crops with profitable ones, matured, and, in fact, over-ripened.

In these terms, the imposed crops of the Culture System sorted themselves out into two broad categories: annuals (sugar, indigo, tobacco), which could be grown on sawahs in rotation with rice; and perennials (coffee, tea, pepper, and less important cinchona and cinnamon) which could not.[11] As a result, these two cultivations developed sharply contrasting modes of interaction with the established biotic communities into which they were projecting, by order of the King. The annuals tended to fall into a mutualistic relationship with such communities, sharing their habitats with them, if not exactly without strain, at least without deterioration on either side and, in fact, with a degree of reciprocal encouragement. The perennials (except for the spices, which had been grown on swiddens since well before the Dutch advent) tended to fall into an insular relationship, pre-empting unused

---

[9] For a history of taxation practices in Java up to 1816, see Bastin, 1957.

[10] Gonggrijp, 1957, pp. 102–103.

[11] This point has been particularly well emphasized by van Klaveren, 1955, pp. 18–19, 118, 120.

habitats and sealing themselves off from the indigenous systems as independent enclaves. It is, to anticipate, but a sort of natural irony, a kind of immanent parable, that, in the long run, it was the mutualistic relationship which turned out to be the more injurious to Indonesian economic vitality.

The two main cultivations, the only ones which occupied much land, absorbed significant quantities of labor, showed important profit, or exercised a lasting influence on the general structure of the peasant economy, were sugar on the annuals side and coffee on the perennials, and they may be taken as type cases.[12]

Sugar demands irrigation (and drainage) and a general environment almost identical to that for wet rice; thus, it was almost of necessity initially cultivated on peasant sawah, for the most part under the one-fifth remission of land-tax procedure.[13] Coffee pre-

---

[12] In 1830, coffee accounted for about 36 percent of Netherlands East Indies exports by value, sugar for about 13 percent; for 1850, the figures are 32 percent and 30 percent; for 1870, 43 percent and 45 percent. Calculated from Furnivall, 1944, pp. 129, 169. For a while, indigo was important, particularly in some dense sawah areas (see, for example, van Doorn, 1926, pp. 37–38). But it never proved very profitable and eventually became almost completely replaced by sugar as the latter flourished. In 1840 indigo accounted for nearly 9 percent of the export value; by 1870 for about 3 percent (Furnivall, as above), and eventually the invention of synthetic dyes removed it from the scene more or less altogether. Coffee and sugar largely dominated the later phases of the Company period too: see Day, 1904, pp. 66–70.

[13] In the indirectly ruled principalities of Surakarta and Jogjakarta, to which this procedure did not formally extend, it was cultivated under the seignorial village-lease system (in which the entrepreneurs rented what were essentially political rights over the villages and the villagers who lived in them from the Javanese aristocracy) which had by now migrated from its original foothold on the northwest coast and shifted from Chinese to European hands. In a few special areas, such as the still sparsely settled Surabaja delta, the quasi-private entrepreneurs rented the necessary sawah and hired the required labor from village authorities on a cash-and-carry basis, usually with capital advanced by the government on condition that the crops produced be sold to it at contracted prices. But these latter systems added up to mere variations on the basic approach—the establishment

fers a highland setting, does not need irrigation, and requires a relatively constant labor force rather than the seasonally variable one of sugar; thus, it was grown on so-called "waste" (that is, uncultiv*ated*, not uncultiv*able*) land, for the most part under the labor-tax procedure. Sugar obligations were measured in terms of units of land per village which had to be devoted to its cultivation; coffee assessments were levied in terms of the number of trees each conscripted family had to care for.[14] It would be expected, then, that sugar, integrated into the sawah regime, would become a peasant crop and coffee, isolated from peasant agriculture, would become an estate crop. Instead, however, in the final three decades of the Colonial Period, about 60 percent of Indonesia's coffee production was coming from (almost entirely Outer Island) small holders, and more than 95 percent of her sugar production (still wholly confined to Java) from Dutch-owned corporate plantations.[15]

This paradox dissolves, however, when the mutualistic and "exclusivistic" relationships are considered. In the mutualistic relationship, the expansion of one side, sugar cultivation, brings with it the expansion of the other, wet-rice growing. The more numerous and the better irrigated the terraces are, the more sugar

---

of a symbiosis between the Javanese subsistence economy and the Dutch commercial economy. For a description of these variant arrangements, see Reinsma, 1955, pp. 125–159.

[14] van Klaaveren, 1955, p. 120. Of course, the relation between labor-force requirements, environmental factors, and the location of the two sorts of cultivation was again a systematic, not a linear one. In part, at least sugar was planted in the lowlands, because that was where the usable population was, coffee in the highlands, because that was where the usable land was. The relative importance of ecological and economic variables in determining the spatial distribution of estate agriculture in Java is difficult to determine with any precision at this late date.

[15] For coffee, Metcalf, 1952, p. 70. Small-holder sugar production, because it was so marginal, is much harder to determine precisely; my 5 percent or so estimate is based on the graph on p. 418, of van de Koppel, 1946. A discussion of small-holder sugar-growing and the reasons for its weakness can be found in van der Kolff, in Ruopp, 1953.

can be grown; and the more people—a seasonal, readily available, resident labor force (a sort of part-time proletariat)—supported by these terraces during the nonsugar portion of the cycle, can grow sugar.

The dynamics on either side of this odd ecological bond support their respective growths. If terracing is improved or extended, because of more and better irrigation, the peasant food production and commercial cultivation can both be increased although they are being grown, so to speak, on the same land. This, in turn, makes it possible to conscript maximum amounts of land and labor into the commercial sector on a temporary basis while leaving that sector "morally" free to contract whenever market conditions so indicate, under the theory that the peasant is only passively engaged in the wider economy and that such contracting merely provides him—until prices recover—with more land to farm and more time to farm it. For men whose whole thought and feeling is consumed by export statistics, as Gonggrijp has said of van den Bosch's, this is an attractive arrangement.[16]

The pleasing symmetry of this picture assumes that population increase is at least matched by the intensive or extensive growth of sawah, and, as we shall see, this eventually came very much not to be the case. It also assumes that the profits to be gained from sugar (or other products) will not prove so fatally attractive as to lead to an overexpansion of its cultivation at the peasant's expense—a danger the government always seems to have realized but not always seems to have been able, or perhaps willing, to avoid. And, more to the immediate point, it assumes that there will be no drift of the market mentality across the export-subsistence line; namely, that Javanese peasants will not themselves replace the cultivation of rice on their lands by small-holder sugar. If they do, the resultant pressure on the subsistence base means that it will be more difficult to conscript peasant land

---

[16] Gonggrijp, 1957, p. 195.

and labor. The workability of the whole mutualistic relationship depends, in short, on each side "doing its job"—the subsistence side feeding the labor force, and the commercial side producing state revenue.

Nor does this change essentially if, as also soon occurred, forced labor is replaced by paid labor, if land is rented rather than its use appropriated as a form of taxation, and if private entrepreneurs replace governmental managers. Then, it is a matter of holding down money wages and rents and avoiding the formation of a true proletariat without the productive means with which to provide its own subsistence.[17] In the Culture System and its appendages political control was pervasive enough, particularly in the settled sawah areas, to prevent the expansion of peasant cane growing: the main problem seems to have been to contain the commercial pressures sufficiently to keep the peasants from fleeing their land altogether. Later, Dutch control of milling, legal restrictions, and semilegal pressures easily effected the same aim of keeping small-holders out of the cane-growing business, and there was no place left to which to flee. In the framework of a colonial political system, the close ecological tie between sugar and rice became the basis for their radical economic separation.

For coffee, the situation was different. Introduced first during the Company period in the Priangan highlands of Sunda under a proto-Culture System arrangement, and later spread to the mountainside areas of the still sparsely settled East Hook, and to some extent of the Kedjawèn, it became by the beginning of the nineteenth century Indonesia's most profitable export. (When the Padri wars gave the Dutch their entree into the Padang highlands after 1837, the forced cultivation of coffee was introduced there too, about the only important example of the Culture-System type of approach outside of Java.) Cultivated, thus, on "waste" land, coffee's fortunes were only indirectly linked with those of peasant subsistence agriculture. Coffee did not demand

---

[17] Even under the Culture System a certain quantity of money wages and rents were paid. See Burger, 1939, pp. 140 ff.

the periodic efforts of great hordes of peasant-coolies organized ant-like into short, intensive "campaigns," but the steady, painstaking application of at least semiskilled labor, and so it was cultivated by a labor force less massive and less fluctuating than the one employed in sugar.

Such a system naturally leads to the formation of enclave estates, manned by permanent, fully-proletarianized workers. Although under the Culture System labor was still drawn almost wholly from the lower-lying rice villages, true plantations, with coolie settlements established on or around them, soon developed, particularly in eastern Java where a supply of landless peasants was available from well-populated but largely sawah-less Madura. By the end of the colonial period, the divergence between this pattern and that characteristic of sugar was virtually complete: of the four sugar enterprises studied by the Coolie Budget Commission Report of 1939–40, none had any resident cane workers; for the three coffee estates, 100 percent of the coolies were housed on and by the estate.[18]

Yet, at the same time, the Dutch were much less concerned to discourage, or prohibit outright, peasant cultivation of coffee than of sugar. A perennial, coffee could not be grown on sawah, and as the government regarded all "waste" lands as its personal property, the reservation of uncultivated land to estate exploitation was easily managed. On the other hand, coffee-growing fitted well with swidden, particularly on hillsides, where small gardens could be cared for without any real pressure on subsistence cultivation and, in fact, as the trees helped with the closed-cover problem, perhaps with some benefit to it. Coffee trees could be planted in swiddens and, taking three or four years before they became productive, could be maintained as gardens after the other swidden crops—grains, legumes, roots— had ceased to be economic and had been "shifted" to another

---

[18] Living Conditions of Plantation Workers and Peasants on Java in 1939–1940, 1956, pp. 6, 13. The distinction is not always quite this sharp, of course, but the contrast is clear over-all.

plot, thus fitting neatly into the phased intercropping pattern of swidden generally. Like pepper before it and rubber after it, coffee could be woven directly into the fabric of swidden farming, and so it spread from its Dutch-established Outer Island beachheads in west Sumatra, north central Celebes, and other areas to free small-holder cultivation in a way sugar, bound to sawah with that dismal combination of dependence and separation we associate with Siamese twins, never did.[19]

This spread, like the full development of enclave plantations, occurred well after the decline of the Culture System, when the Outer Islands began to feel more directly the impact of Western commercial interests. But the basis for the spread was laid during the decline. The isolation of coffee cultivation on enclave European estates, like that of rubber later on, made the barrier to the drift of a commercial orientation to agriculture into the peasant sector much less formidable in the long run than did the mutualistic integration of sugar with wet-rice growing. Here, ecological separation eventually reduced economic contrast, at least in the sense that peasant agriculture became a functioning element in the Indies' export economy rather than merely its backstop; peasant agriculture was developed, at least in part, into a business proposition rather than becoming frozen into a kind of outdoor relief. By the end of the colonial period, around 45 percent by value of Outer Island peasant agriculture was embodied in export crops as against food crops, in Java about 9 percent. Put another

---

[19] For a discussion of the role of tree crops in the swidden cycle and their continued harvesting over extensive periods, see Conklin, 1960. For the integration of coffee trees into the swidden pattern in Outer Indonesia, see van Hall, n.d., pp. 105–106; Paerels, 1946; and Pelzer, 1945, pp. 23–26. For pepper, see Rutgers, 1946. For rubber, van Gelder, 1946; and Thomas, n.d., pp. 21–23 and Appendix H. Tree crop planting in swiddens is, of course, not free of the adaptive constraints of the regime generally; much of southwest Sumatra was planted with *Robusta* coffee trees in the early 1930's, only to be abandoned with the collapse of coffee prices in the depression, leaving vast areas to revert to *imperata* savannah (Benjamin Higgins, personal communication).

way: though the production (by value) of food crops per capita was approximately the same, the per-capita value of export crops was about seven times as high in the Outer Islands.[20]

But despite this difference in relationship to existing ecological patterns, annuals and perennials grown under Dutch auspices shared another critical property that peasant cultivations (even when they were the same crops and were grown for the same commercial ends) lacked—they were integrated into a modernizing economy.

This basic difference—a sociological and not an ecological one now—split Netherlands East Indies agriculture into two radically unconformable strata which, despite Boeke's objections, one can only call "native" and "foreign," or, perhaps better, Indonesian and Dutch.[21] For the large-scale, well-capitalized, rationally organized estate agriculture which by 1900 accounted for 90 percent by value of Indonesia's exports (by 1938, after Outer Island small holders got well established, 60 percent), was essentially not part, save in a merely spatial or geographic sense, of the Indonesian economy at all, but of the Dutch.[22] The universal practice of colonial historians of opposing the NEI economy as a unit on the one hand to the Netherlands economy as a unit on the other merely obscures this crucial fact. There never really was, even in Company times, a Netherlands East Indies economy in an integral, analytic sense—there was just that, admittedly highly autonomous, branch of the Dutch economy which was situated in the Indies ("tropical Holland," as it was some-

---

[20] Boeke, 1947, p. 25. As Boeke remarks, such figures (which are for 1939 and are computed on the agrarian population alone) can be at best general estimates, particularly on the food crop side, but the order of magnitudes is such that the broad picture is clear.

[21] Some of the "Dutch" side actually came into British, American, etc., hands; by 1937 about a quarter (Allen and Donnithorne, 1957, p. 288). Boeke's desperate attempt to dissociate dualism and colonialism is to be found in his 1953, pp. 18–20 and *passim*.

[22] Metcalf, 1952, p. 8.

times called), and, cheek-by-jowl, the autonomous Indonesian economy also situated there. And though, indeed, the two interacted continuously in ways which fundamentally shaped their separate courses, they steadily diverged, largely as a result of this interaction, to the point where the structural contrasts between them were overwhelming. What Boeke regarded as an intrinsic and permanent characteristic of Indonesian (or "Eastern") economic life, "a primarily spiritual phenomenon," was really an historically created condition; it grew not from the immutable essence of the Eastern soul as it encountered the incarnate spirit of Western dynamism, but from the in no way predestined shape of colonial policy as it impressed itself upon the traditional pattern of Indonesian agriculture.[23]

This view, that the radical economic dualism which came to characterize Indonesia was brought on by a set of colonial policies may seem to be belied by the fact that it is found in countries which are not colonies (Italy, for example) and also may persist in former colonies (the Philippines, for example) after they have become independent. But the fact that a phenomenon is general does not mean that the particular occasions of its appearance may not be various, as the example of inflation only too well demonstrates. Basically, the development of dualism consists of a trend toward fixed (or presumed fixed) technical coefficients of production in more and more capital-intensive enterprises on the one hand, and toward variable ones in more and more labor-intensive activities on the other, together with the peculiarly lopsided pattern of investment, productivity, and employment which flows from this steadily widening disparity.[24] Despite its formal identity from place to place, the ways in which such a situation can arise historically are very diverse. And as what we are concerned to understand here is not the mere presence but the extraordinary severity of this development so far as Indonesia is concerned (with, so to speak, "runaway dualism"),

[23] Boeke, 1953, p. 14.
[24] Higgins, 1959, pp. 325–340.

the nature of Dutch colonial policy, the decisive force on the capital-intensive side, can hardly be irrelevant.

The inability of Dutch private enterprise to provide the capital necessary to exploit Java efficiently was one of the main motivating forces for the institution of the Culture System in the first place.[25] From 1816 (when the English interregnum ended) to 1830 (when van den Bosch, plan in hand, arrived in Java), the Dutch in Indonesia faced a situation similar to that faced by some newly independent nations today. A once effective mechanism for producing foreign exchange—the Company—had become exhausted and discredited, and had disappeared, leaving behind it an intense theoretical controversy over how to increase the island's profitability. Land and labor were available enough, but capital was in short supply, which hampered rapid expansion of private enterprise (at least by the Dutch) of the sort desired by the Liberal opponents of van den Bosch.[26] Looked at against this background, the Culture System appears to represent the kind of governmental mobilization of "redundant" labor for capital creation projects which has been often proposed and occasionally attempted in underdeveloped areas. Within the framework of the labor-tax system, itself cast in the mold of the traditional corvee powers of the indigenous aristocracy, the government built roads and bridges, expanded irrigation facilities, cleared and improved large tracts of "waste" land, constructed buildings, and generally substituted the labor of the Javanese for the capital Holland lacked in laying the preparative foundations of a very rapidly accelerating, if distorted, process of economic growth.

At first, such efforts to accumulate social capital by applying redundant labor to government projects were, like the forced work in cultivation itself, not altogether successful. (Nor was the

---

[25] Reinsma, 1955, pp. 17–21.

[26] For a description and analysis of the Liberal-Conservative debate of this period (and a spirited, not to say emotional, defense of the Conservative policy actually chosen), see Ottow, 1937.

labor always so simply redundant as it seemed, as the rice-crop failures and attendant famines of the 1840's soon made clear.) An estimated 100,000 days of unpaid labor on an irrigation system in Demak went largely, and literally, down the drain, when typical northcoast wet monsoon flooding washed out its main work.[27] In Tjilatjap, on the south coast, even more work seems to have been wasted when two large-scale attempts to build a waterway to the harbor failed, one because an impossible route had been chosen, the other because the banks caved in.[28]

But, as in the cultivations, experience proved a teacher, and these projects soon grew more efficient and more extensive. In the so-called Bagalen area west of Jogjakarta in the Kedjawèn a huge irrigation canal was built to support the forced indigo cultivation and to provide water for the new "capital" of the region of Purweredjo; the lower reaches of the Lukolo river were rerouted to run more directly to the sea, thus draining a good part of the coastal swamp which had made the littoral uncultivable up to that time; and four large, as well as a great number of small, indigo factories were built with taxation labor.[29] Clive Day, drawing on a British government source, tells of 1,200 men, cheered by musicians and dancing girls, laboring three months on a single dam somewhere in Java during this period; and in the Brantas delta around Sidoardjo, "private" entrepreneurs, using government-supplied labor, began to build up what eventually became, after 1852, Indonesia's first technically modernized irrigation system.[30] Concerning roadways and bridges, even fewer specific data seem to be available, but Furnivall remarks that in the beginning of the Culture System (when there was but a single trunk road to the interior), communications threatened

---

[27] Gonggrijp, 1957, p. 103.

[28] Day, 1904, p. 286. A third effort ultimately succeeded, however.

[29] Van Doorn, 1926, pp. 37–38. There was also a plan to build three sugar mills which never materialized. Elsewhere in Java, however, government labor was directed to mill construction: see Reinsma, 1955, p. 138.

[30] Day, 1904, pp. 285–286.

to be a greater bottleneck to Java's development than its fertility:

. . . but the administrative machinery of van den Bosch gave the State an almost unlimited command over free labor, and this was employed so lavishly in some parts, if not everywhere, that in 1847 van Hoevell talks of the fine roads intersecting the countryside in the Preangers, with bridges spanning the numerous streams and of the imposing public buildings; and a few years later Money was impressed by the superiority of the communications to those of British India.[31]

From the developmental point of view, therefore, the Culture System represented an attempt to raise an estate economy by a peasantry's bootstraps; and in this it was remarkably successful. Benefiting from the external economies created by the formation of social capital, the forced diffusion of plantation crops and attendant labor skills over the island, and a certain amount of more direct governmental assistance, private enterprises steadily multiplied; soon their returns were great enough that they could provide most of the investment required for the qualitative changes in capital stock, particularly in sugar-milling, which were becoming necessary.[32] As Reinsma has well argued, the protracted "fall" of the Culture System (which lasted from about 1850 to about 1915) and its gradual replacement by the Corporate Plantation System were largely self-generated, because its success in establishing a serviceable export economy infrastructure made private entrepreneurship, originally so hampered by lack of capital, progressively more feasible:

---

[31] Furnivall, 1944, p. 128. Van Hoevell was an archenemy of the system, Money a British enthusiast for it.

[32] In 1840, private estates accounted for 17 percent of agricultural export volume, government forced cultivation 78 percent. In 1850, the figures were 26 and 73; in 1860, 58 and 39; in 1870, 43 and 52; in 1873, 72 and 19. Reinsma, 1955, p. 157. For a general discussion of "social overhead capital," "external economies," and development, see Higgins, 1959, pp. 384–408.

A great many [private Netherlands East Indies] planters . . . had their success to thank not to the [economic] renaissance of the Netherlands, but to "the energy of pioneers, supported by speculative elements in the Indies itself." So far as capital supply is concerned, private enterprise in the motherland played a much less powerful role in supporting the successors of the Culture System than has commonly been suggested in the literature.[33]

Like most Liberal analyses, this gives more weight to individual character and less to social context than seems realistic. But its essential point is accurate: though importantly assisted by Netherlands investment in shipping and trade, the capital-intensive side of the Indonesian dual economy was not merely imported bodily from Holland like street canals and billiards, but was a direct product of the workings of the Colonial System after about 1830.

If we look again at the two leading commercial crops of the nineteenth century, coffee and sugar, a somewhat more palpable picture of the developmental pattern appears, which the Culture System "big push" generated. (Graph 1.)[34] The production of coffee, the cultivation most immediately affected by the system, rose sharply within ten years of the inception of mass-labor taxation to a new level as it spread extensively over the uncultivated uplands of Java. (In 1833 somewhat more than 100 million trees are said to have existed on Java; two years later, in 1835, about twice that many, and by 1840–50 more than three times as many.[35]) But, not particularly susceptible to qualitative technical

---

[33] Reinsma, 1955, p. 171. The internal quotation is from a Dutch government report on "the first commercial houses in Indonesia."

[34] Based on figures in Furnivall, 1944, pp. 75, 104, 129, 171, 207. The beginning and end points of the Culture System are the conventional ones, and admittedly somewhat arbitrary. No production figures for 1870 are available, evidently.

[35] Van Klaaveren, 1955, p. 123. As not even Dutch civil servants can have been so conscientious as to have counted 300 million coffee trees, such figures must, of course, be taken only as general, order-of-magnitude estimates.

GRAPH I

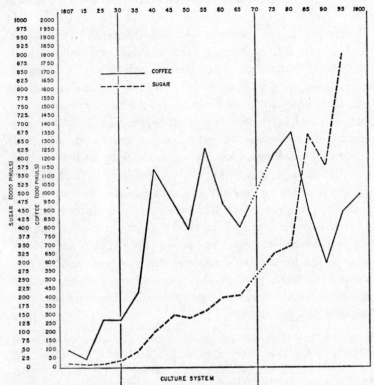

SOURCE: J. S. Furnivall, *Netherlands India,* Cambridge (England): Cambridge University Press, 1944, pp. 75, 109, 129, 171, 207.

improvement, this spread did not continue upward once it reached its approximate limits; having found a higher level, it oscillated around it. (The severe dip in 1880–1890 was due to disease, which was overcome in the 1890's by a shift in species.) For the period of the Culture System proper, roughly the middle half of the nineteenth century, coffee was to Java what textiles were to England—it virtually carried the estate economy, accounting in itself for between a quarter and a third of the Indies' export income.[36]

[36] Furnivall, 1944, as in note 34. Though price movements obviously shaped this pattern as it developed, they can in no simple way account for it. From 1825 to 1855, the general trend of coffee prices was moderately

Sugar production, however, behaved differently. In the first place, it rose less explosively (and also less erratically) under the Culture System, mainly because its advances were based on technical improvements, in both cultivation and milling, which took time and capital to develop. Where the area planted in cane increased only about 18 percent between 1833 and 1861, the production per hectare of refined sugar about tripled.[37] And, unlike the coffee expansion, this growth, once underway, continued at an accelerating rate: by 1900, productivity had doubled again and, now that modern irrigation facilities were appearing in significant quantity, planted area had about tripled, leading to the five-fold output increase shown on the graph.[38] Thus, the role of vanguard industry that coffee played up to about 1880, sugar played for the last quarter of the nineteenth century and into the twentieth, until in 1920 it earned more than a million guilders of export revenue—more than all other products, minerals included, put together.[39]

---

downward; after 1860 they rose rapidly until 1875, when they plunged downward again, rising once more in 1890–1895 and falling off in 1900 (calculated from Furnivall). Sugar prices were somewhat less fluctuating, at least until 1885, but a similar analysis can be given for them.

[37] Burger, 1939, pp. 130, 154. On the cultivation side the technical improvements consisted of more intensive planting and care under the so-called Reynoso system; on the milling side of more and more elaborate machinery. Between 1830 and 1839 machinery imports (mostly for sugar) into the Netherlands East Indies averaged 1.48 million guilders annually; 1840–1849, 1.79; 1850–1859, 2.02; 1860–1869, 2.78 (Reinsma, 1955, p. 159). Capitalization was increasing in other crops too: after 1859, the private tobacco grower George Birnie spent more than 500,000 guilders on irrigation works in Djember (Boeke, 1953, pp. 212–213.).

[38] Gonggrijp, 1957, p. 121. Gonggrijp's figures do not entirely tally with Burger's, but again the general direction, if not the precise values, is clear enough. For the steady increase in per-hectare refined sugar productivity from about 20 quintals in 1842 to about 165 in 1937, see Koningsberger, in van Hall and van de Koppel, 1946, p. 291.

[39] Furnivall, 1944, p. 337. About this time Java was contributing about 10 percent of the world's sugar supply; see graph in Bulletin of the Colonial Institute of Amsterdam, 3:207 (1939).

In sum, if "take-off" is defined as a largely self-generated, relatively sudden transition to sustained economic growth, then there is at least presumptive evidence that something of the sort occurred on the estate side of the dual economy during the Culture System period, or—if the system itself be viewed as merely establishing preconditions—just after it.[40] The fact that the estate sector became progressively more closely integrated into the modernizing Netherlands economy (and progressively segregated from the rigidifying Javanese one) does not mean, as has sometimes been asserted, that it was a simple creation of that economy; it was a creation of Javanese land and labor organized under Dutch colonial political direction. If anything, the flow of support ran the other way: "The true measure of [van den Bosch's] greatness," Furnivall has justly written, "is the renascence of the Netherlands." [41]

The true measure of van den Bosch's malignancy, however, is the stultification of Indonesia. For, although the Javanese helped launch the estate sector, they were not properly part of it, nor were they permitted to become so; it was just something they did, or more exactly were obligated to do, in their spare time. On their own time, they multiplied; and "take-off" on the peasant side was of a less remunerative sort—into rapid and sustained population growth. In 1830 there were probably about 7 million people on Java; in 1840, 8.7; 1850 (when census-taking first became reasonably systematic), 9.6; 1860, 12.7; 1870, 16.2; 1880, 19.5; 1890, 23.6; 1900, 28.4—an average annual increase of approximately 2 percent during seventy years.[42] And, here too, the pattern once

---

[40] The "take-off" concept is from Rostow, 1960, pp. 36–58. However, Rostow's analysis generally ignores the developmental patterns of colonial economies.

[41] Furnivall, 1944, p. 152.

[42] Reinsma, 1955, p. 175; The Population of Indonesia, p. 22; van Alphen, 1870, second part, p. 68. Not all these sources precisely agree, but the general pattern is about the same. Also, of course, the rate of growth was not steady: in the 1840's when there was famine, the annual increase seems to

established, persists (though the rate slows): 1920, 34.4; 1930, 41.7. What the precise causes of this "explosion" were and, particularly, how far it was directly rather than merely indirectly detonated by Culture System policies, are, reliable data being scarce, matters of debate. But there is little doubt that it was during the Culture System period that the saying about the Dutch growing in wealth and the Javanese in numbers first hardened into a sociological reality. By the end of it, the Javanese had, as they have today, the worst of two possible worlds: a static economy and a burgeoning population.

The unreliability (as well as the scarcity) of agricultural statistics on the peasant sector during the nineteenth century, on both the extent of cultivated land and the production of food crops, makes an exact tracing of the way in which the Javanese coped with this deepening demographic dilemma difficult. But, although the stages through which their adaptation passed have to be described in speculative terms, shored up only by fragmentary and indirect evidence plus some hard reasoning, the over-all nature and direction of that adaptation are clear, and comprise what I am going to call "agricultural involution."

Unable to attack the problem head-on, let us attempt to get at it by a circuitous route. Beginning with a picture of the general situation at a later period, when statistics on peasant agriculture are more reliable, we can try, first, to figure out how the situation characteristic of this later period could have been produced and then, second, we can see whether the scattered historical evidence supports the notion that it was in fact so produced. Such a procedure amounts, admittedly, to doing history backwards. But it is doing history, not deducing logically the past from the present. It is moving from a known result to an analysis of a factually much less fully outlined process which seems to

---

have sunk to around 1 percent, in the 1850's to have risen to about 3 per cent—though it is difficult to tell how much of such variation stems from irregularities in census-taking. For a review of the development of population surveys in Indonesia, see van de Graaf, 1955, pp. 138–169.

have brought the result about, in order to clarify that process and give it a more concrete content.

In any case, the striking set of figures with which I want to

TABLE 1

THE LAND, POPULATION, AND RICE-PRODUCTION CHARACTERISTICS
OF THE SUGAR REGIONS OF JAVA IN 1920

	Percent of land of Java	Percent of wet-rice land of Java	Percent of popu- lation of Java	Percent of wet-rice pro- duction of Java
The 37 main sugar regencies (47% of the total number of regencies)	34	46	50	49
The 98 main sugar districts (22% of the total number of districts)	15	22	24	24
The 19 leading sugar districts (4% of the total number of districts)	2.6	4.6	5.3	5.2
	Same with Percentage of Land Set as Index Base			
The 37 main sugar regencies	100	131	143	140
The 98 main sugar districts	100	147	160	160
The 19 leading sugar districts	100	178	204	200

SOURCE: Calculated from data in: Landbouwatlas van Java en Madoera, Mededeelingen van het Centraal Kantoor voor de Statistiek, No. 33, Weltevreden: 1926, Part II, Tables I, III, IV, and V. In Mataram (which in 1920 was organized slightly differently from the rest of Java), Bantul, Sléman, and Kalasan have been counted as districts, Mataram itself as a regency.

TABLE 2

RELATIONSHIPS BETWEEN WET-RICE PRODUCTION, SAWAH AREA,
HARVESTED SAWAH AREA, PER-HECTARE WET-RICE YIELDS AND
POPULATION IN THE SUGAR AREAS OF JAVA IN 1920
(Indexes only, with all-Java figure set as base)

	Ratio: Percent of wet-rice production/ percent of sawah area	Ratio: Percent of wet-rice production/ percent of population	Percent of sawah area harvested in rice (includes double cropping)	Average per-hectare yields of harvested (in rice) sawah
The 37 main sugar regencies	107	98	98	110
The 98 main sugar districts	109	100	94	115
The 19 leading sugar districts	113	98	85	128
All-Java (base)	100	100	100	100

SOURCE: Same as Table 1.

begin are those of Tables 1 and 2, which show the character of
the sugar areas of Java set against the population and rice-production characteristics of the island as a whole in the year that
probably was the high-water mark of the colonial economy of
Indonesia—1920. Moving from the level of the larger and more
inclusive administrative unit, the regency, to the smaller, constituent unit, the district, and then to selected districts, the tables
focus on the sugar areas of Java.[43]

The "thirty-seven main sugar regencies" are those in which
there were, in 1920, at least a thousand hectares planted in sugar.

_____

[43] The order of territorial administrative units in Java is: province, residency, regency, district, subdistrict. On the average there were slightly less
than six districts per regency in 1920.

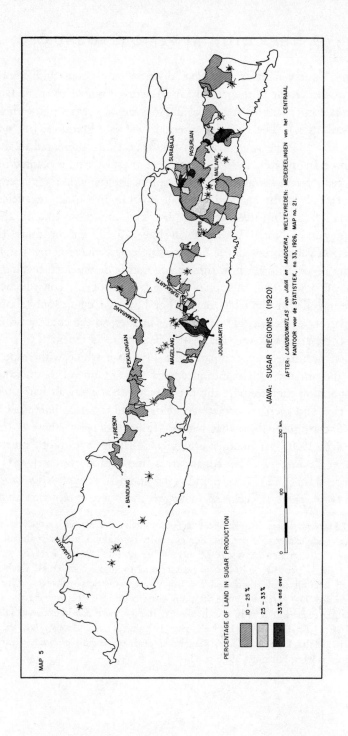

MAP 5

PERCENTAGE OF LAND IN SUGAR PRODUCTION

10 – 25 %

25 – 33 %

33% and over

JAVA: SUGAR REGIONS (1920)

AFTER: *LANDBOUWATLAS van JAVA en MADOERA*, WELTEVREDEN: MEDEDEELINGEN van het CENTRAAL
KANTOOR voor de STATISTIEK, no 33, 1926, MAP no. 21.

DJAKARTA

BANDUNG

TJIREBON

PEKALONGAN

SEMARANG

MAGELANG

SURAKARTA

JOGJAKARTA

KEDIRI

MALANG

PASURUAN

SURABAJA

0   100   200 km.

The "ninety-eight main sugar districts" are those with more than 10 percent of their total sawah acreage rented to sugar factories for cane growing in a single year (see Map 5 for their distribution). The shift in criterion here is regrettable, but necessitated by the way in which the districts are presented in the source. In any case, as about 94 percent of estate cane is planted on rented peasant sawahs, the correlation of the two criteria is nearly exact: all of the main districts lie within the main regencies. Finally, the "nineteen leading districts" are those among the ninety-eight "main" ones which have 25 percent or more of their sawah so rented, the highest single percentage being a startling 44 percent. Each of these categories is demarcated by sharp cut-off points, and, therefore, all sugar areas of any consequence are included in the table totals. (For Java as a whole, only about 8 percent of her sawah was in cane in 1920, so the concentration in these areas was very sharp.) Table 2 presents indexes showing how the three units compare with Java as a whole with respect to the amount of rice produced for the amount of sawah and population they contain, the percentage of their sawah harvested in rice, and average per hectare yields of the sawah so harvested.[44]

These figures show that within "Inner Indonesia" there is (or was in 1920) yet another ecological nucleus, a sort of "inner" Inner Indonesia. The sugar areas have proportionately: (1) more sawah; (2) more population; and (3) *even though more of their sawah is occupied by sugar,* more rice production than

---

[44] Harvested area includes double (occasionally even triple) cropping, so totals may exceed 100 percent. For example, Banjudono, a district in Surakarta, managed to harvest 147.5 percent of its total sawah acreage, even though 26 percent of its sawah was rented to sugar companies! Such a high over-all total is unusual for heavy sugar districts, however, which averaged 90.4 percent for the 98 main districts and 81.4 percent for the 18 leading ones. But if only sawah planted in rice were considered, virtually all such districts would be well over 100 percent. Landbouwatlas, 1926, as in Tables I and II. The 8 percent of Javanese sawahs planted in cane figure is from Huender, 1921, p. 83.

the nonsugar areas. Further, the relationship strengthens markedly, as the sugar areas are more precisely pinpointed by isolation of the cane districts within the sugar regencies generally and the leading cane districts among all such districts. Whatever the causes, the tie between sugar, wet-rice, and population density is unmistakable: all three "flourish," if that is the proper word, together.

In Table 2, the main factor which makes this seemingly contradictory phenomenon possible is revealed: a progressively higher *per hectare* sawah productivity. Taking Java-wide averages as a base, yields can be seen to rise as one moves to more intensely planted sugar areas, a rise which, especially when combined with the concurrent (but slower) rise in sawah, compensates for the loss of rice land to cane. This, in turn, brings about the more moderate rise in wet-rice production per total sawah area. The question, then, is: why the higher yields?

Two possibilities suggest themselves: (1) sugar tended to be planted in the best rice areas; (2) rice cultivation was more efficiently pursued in the sugar areas. Both of these factors certainly were operative. Given the ecologically specialized, "aquarium" nature of wet-rice growing, it is difficult to distinguish sharply between them because fertility is so closely tied to level of technique rather than merely reflecting natural conditions; the best rice areas are naturally the better-worked areas, and vice versa. As already explained, the ecological requirements for cane and rice are similar, therefore sugar gravitated to the most fertile (i.e., the best-irrigated) sawahs and, by financing water-control facilities, the companies improved and expanded such areas.

This thesis that sugar cultivation (and to a lesser extent other mutualistic plantation cultivations) was an operative, not merely an accidental or derivative, link between high density, high "sawah-ization," and high productivity is strengthened by the following argument. First, though on the whole sugar districts tended to be dense (and the densest districts tended to be sugar districts), not all dense districts were also sugar districts.[45] Second, though sugar districts tended to be productive rice districts, not all of the most productive rice districts were also cane districts (though if the sugar industry had continued to expand, perhaps nearly all would eventually have become so).[46] Third, a similar proposition holds for percentage of total arable land irrigated.[47] Fourth, and finally, if sugar districts are then eliminated from consideration, the close correlation between per-hectare wet-rice yields, population density, and percentage of arable land which is irrigated disappears or at least is noticeably weakened. As Kuperus has shown, for the universe of *nonsugar* districts (what he calls "districts in the indigenous sphere") the highest-yield districts are *not* those with the densest population,

---

[45] Excluding the heavy urban concentrations (Djakarta, Bandung, Surakarta, Jogjakarta, Semarang, and Surabaja), of the 27 districts which had more than 700 persons per square kilometer in 1920, 20 were "main" sugar districts; of the 78 which had more than 500, 49 were "main" sugar districts. Landbouwatlas, Vol. II, "Text and Tables," Table I.

[46] Of the 77 districts in which rice yields were more than 20 percent above the Java-wide average, 38 were "main" sugar districts, 39 were not. Landbouwatlas, Table III. The bulk of the latter were, however, contiguous to sugar districts and so were subject to some of the same influences. Of the 121 districts with yields 20 percent below the Java average, only three were "main" sugar districts, and they were just barely below the line.

[47] For the 37 main sugar regencies, the percentage of total cultivated land which was irrigated was about 50, for the 98 main sugar districts 56, for the 19 leading districts, 64; for all Java, 45. Landbouwatlas, Table I. Actually, the index problem is complicated here by the fact that many less developed districts, especially in west Java, without much cultivated land of any kind have what little they do have in sawah, while others, especially in east Java, have what little they do have in dry fields, so that an aggregate distribution by percent of land irrigated gives a misleading and incomprehensible picture. If one could group villages into subtypes according to general level of agricultural development and then compare within subtypes, the picture would be clear and systematic; but this is difficult to do in a nonarbitrary way on the basis of the Landbouwatlas data.

those with the highest sawah proportions are not the densest, nor those with the highest sawah proportions the most productive.[48] Evidently, sugar cultivation, through its improvement of local ecological conditions for rice, binds those three together when they are found together, and pushed all of them to the higher than average levels shown above.[49]

But without more effective cultivation methods the Javanese could hardly have taken advantage of these better facilities. And as there was virtually no variation in capital inputs in sawah agriculture from one part of the island to another, aside from irrigation works, this greater efficiency in cultivation derived almost entirely from a greater intensification of labor—an intensification made both possible and necessary by the increasing population. The practices have already been mentioned—pregermination, transplanting, more thorough land preparation, fastidious planting and weeding, razor-blade harvesting, double-cropping, a more exact regulation of terrace-flooding, and the

---

[48] Kuperus, 1930.

[49] Kuperus, who is concerned to show that Java's failure to progress is due not to colonial impact but to internal stagnation of Javanese "culture" since classical times, has tried to deny a link between rice yields, Western irrigation methods, and population density for sugar areas as well. ("The Javanese culture . . . since the fall of Madjapahit has been a *fellah*-culture. The Javanese people have had their culture and try now only to hang on to what cultural resources they once had. Javanese culture has stagnated and the problem of population pressure is therefore more closely connected to the ancient times than to the past one hundred years of Europeanization.") Kuperus, 1930, and 1938. But as this conclusion is based on the study of three closely grouped districts in the Tjireborn-Pekalongan northcoast plain, two of which happen to be among the least productive of the sugar districts, the argument is unconvincing. Only when one takes the whole universe of sugar districts does the relative importance of various determinants and their interactions emerge. By selecting an area in which, evidently for local ecological reasons (abundance of water together with distance from sources of volcanic fertilization), expansion of percentage of land in sawah has played a greater role than increasing yields, Kuperus got a reversed picture of their general relationship.

addition of more fields at the edges of volcanoes.[50] The concentrative, inflatable quality of sawah, its labor-absorbing capacity, was an almost ideal (in an ecological, not a social, sense) complement to capital-intensive sugar-growing. As Table 2 shows, it enabled the densest regions of Java to keep pace (at least until 1920) with the per-capita output of rice of the island as a whole.

The process resembles nothing else so much as treading water. Higher-level densities are offset by greater labor inputs into the same productive system, but output per head (or per mouth) remains more or less constant from region to region. This, however, is only the synchronic picture. From the diachronic point of view, the important questions are: how long has this water-treading been going on? What set it off? What sustained it? And in this respect, two otherwise isolated facts take on significance: first, that *local* overcrowding begins to be reported from Java as early as the beginning of the nineteenth century, when the island had a density about like that of Thailand today;[51] second, that, with the exception of a disproportionate expansion in the Brantas basin, due mainly to large-scale irrigation improvements carried out in the 1890's, the distribution of sugar cultivation in 1860, and for the most part even in 1833, was about the same as it was in 1920, though only about a sixth as much acreage was involved.[52]

---

[50] Though they tend to increase labor intensification, such procedures as double-cropping or pushing terraces up mountainsides do not, of course, necessarily increase per-harvested-hectare yields, but increase production mainly by increasing harvested area.

[51] Boeke, 1953, pp. 166–167. In 1817 Raffles remarks that "the population of Java is very unequally distributed," and that the bulk of the rice production comes from about one-eighth of the island. Raffles, 1830, I, 68, 71, 119. On the basis of colonial reports Kuperus (1930) estimates a nutritional density for 1827 Java of .35 ha. cultivated land per capita. In the "mid-1950's" Burma's nutritional density was about .44, Thailand's .38, Malaya's .31, the Philippines' .27 (Java, 1956, .15). Ginsburg, 1961, p. 46; Statistical Pocketbook of Indonesia, 1957, p. 46.

[52] Burger, 1939, p. 132.

These fragments suggest that the ever-more energetic, regionally uneven process of water-treading which led to the swollen population and intensely driven rice terraces of 1920 and later went on steadily during the whole of the past century. From 1850 on there are even some reasonably reliable figures to support this view: the per-capita production of rice between 1850 and 1900, when population was mounting as rapidly as it probably ever has in Java, is given as averaging about 106 kilograms, with no clear trend in any direction: 1850, 106; 1865, 97 (the low); 1885, 119 (the high); 1895, 105; 1900, 98, and so on.[53] After the turn of the century the 1900–1940 average declined to about 96.[54] But, as we shall see, by then the expanded cultivation of dry crops had begun to take up the slack and total per-capita output probably remained at about the nineteenth-century level. In a general review of the problem, Hollinger has written: "Taking all the historical evidence available into consideration, we conclude that per capita food consumption has been maintained through the period of rapid population increase, but it has never risen above a minimal level." [55] Less circumspect, Boeke summed the whole picture up in a single, mordant phrase: "static expansion." [56]

The superimposition of sugar cultivation on the already unequal distribution of sawah and population over Java left the Javanese peasantry with essentially a single choice in coping with their rising numbers: driving their terraces, and in fact all their agricultural resources, harder by working them more carefully. There was no industrial sector into which to move and, as the returns from cultivation went, in Furnivall's words, to keep the

---

[53] Table quoted from A. M. P. A. Scheltema, 1936, Table II.10, in Hollinger, 1953b. The maintenance of the level after about 1880 seems to have been due not only to increasing yields but new terrace construction made possible by expanded irrigation systems, particularly in the northwestern and eastern parts of the island.

[54] Hollinger, 1953b.

[55] Hollinger, 1953a. See also, Wertheim and The Siauw Giap, 1962.

[56] Boeke, 1953, p. 174.

Netherlands from becoming another Portugal, none was developed.[57] Coffee-growing was still almost wholly a forced-labor occupation and no real substitute for subsistence cultivation; and the same was true of the other Culture System crops. The Javanese could not themselves become part of the estate economy, and they could not transform their general pattern of already intensive farming in an extensive direction, for they lacked capital, had no way to shuck off excess labor, and were administratively barred from the bulk of their own frontier, the so-called "waste lands" which were filling up with coffee trees. Slowly, steadily, relentlessly, they were forced into a more and more labor-stuffed sawah pattern of the sort the 1920 figures show: tremendous populations absorbed on minuscule rice farms, particularly in areas where sugar cultivation led to improved irrigation; consequent rises in per-hectare productivity; and, with the assistance after about 1900 of an expansion in dry-crop cultivation, a probably largely stable, or very gradually rising, standard of living. Wet-rice cultivation, with its extraordinary ability to maintain levels of marginal labor productivity by always managing to work one more man in without a serious fall in per-capita income, soaked up almost the whole of the additional population that Western intrusion created,[58] at least indirectly. It is this ultimately self-defeating process that I have proposed to call "agricultural involution."

I take the concept of "involution" from the American anthropologist Alexander Goldenweiser, who devised it to describe those

---

[57] Furnivall, 1944, p. 151.

[58] Just what the factors producing the population rise in the nineteenth century were is not quite so clear as the usual references to the removal of Malthus' positive checks would make it seem. Improved hygiene could hardly have played a major role until fairly late. The Pax Nederlandica had perhaps more effect, but probably not so much because so many people had been killed in wars in the precolonial period but because the attendant destruction of crops ceased. Probably most important, and least discussed, was the expansion of the transport network which prevented local crop failures from turning into famines.

culture patterns which, after having reached what would seem to be a definitive form, nonetheless fail either to stabilize or transform themselves into a new pattern but rather continue to develop by becoming internally more complicated:

The application of the pattern concept to a cultural feature in the process of development provides . . . a way of explaining one peculiarity of primitive cultures. The primary effect of pattern is . . . to check development, or at least to limit it. As soon as the pattern form is reached further change is inhibited by the tenacity of the pattern. . . . But there are also instances where pattern merely sets a limit, a frame . . . within which further change is permitted if not invited. Take, for instance, the decorative art of the Maori, distinguished by its complexity, elaborateness, and the extent to which the entire decorated object is pervaded by the decoration. On analysis the unit elements of the design are found to be few in number; in some instances, in fact, the complex design is brought about through a multiplicity of spatial arrangements of one and the same unit. What we have here is pattern plus continued development. The pattern precludes the use of another unit or units, but it is not inimical to play within the unit or units. The inevitable result is progressive complication, a variety within uniformity, virtuosity within monotony. This is *involution*. A parallel instance . . . is provided by what is called ornateness in art, as in the late Gothic. The basic forms of art have reached finality, the structural features are fixed beyond variation, inventive originality is exhausted. Still, development goes on. Being hemmed in on all sides by a crystallized pattern, it takes the function of elaborateness. Expansive creativeness having dried up at the source, a special kind of virtuosity takes its place, a sort of technical hairsplitting. . . . Anyone familiar with primitive cultures will think of similar instances in other cultural domains.[59]

From the point of view of general theory, there is much misplaced concreteness in this formulation; but for our purposes

[59] Goldenweiser, 1936. As his own reference to late Gothic art demonstrates, however, there is nothing particularly "primitive" about this process.

we want only the analytic concept—that of the overdriving of an established form in such a way that it becomes rigid through an inward overelaboration of detail—not the hazy cultural vitalism in which it is here embedded.

The general earmarks of involution that Goldenweiser lists for aesthetic phenomena characterized the development of the sawah system after about the middle of the nineteenth century: increasing tenacity of basic pattern; internal elaboration and ornateness; technical hairsplitting, and unending virtuosity. And this "late Gothic" quality of agriculture increasingly pervaded the whole rural economy: tenure systems grew more intricate; tenancy relationships more complicated; cooperative labor arrangements more complex—all in an effort to provide everyone with some niche, however small, in the over-all system. If the original establishment of terraces in Java's little interior river galleries was but a preliminary sketch of the wet-rice mode of adaptation, and the time of the Javanese states and of the Company saw a filling in of solid compositional substance, the Culture System period brought an overornamentation, a Gothic elaboration of technical and organizational detail. But what makes this development tragic rather than merely decadent is that around 1830 the Javanese (and, thus, the Indonesian) economy could have made the transition to modernism, never a painless experience, with more ease than it can do today. To see how this is so, however, it is necessary to look at the last major colonial device for exploiting the archipelago, the Corporate Plantation System, for it is under its aegis that all the immobilizing processes which the Culture System so powerfully propelled settled into their definitive form.

5. THE COLONIAL PERIOD: FLORESCENCE

The Corporate Plantation System

The rapid mechanization of sugar-milling in the second half of the nineteenth century steadily made obsolete the Culture System, based on the substitution of (Javanese) labor for (Dutch) capital. With such substitution rendered progressively less practical by technological advance, effective colonial management became less a matter of mobilization of labor and more of regulating the relationship between the highly capitalized sugar "factory," or other crop-processing enterprises, and the peasant village to which it was symbiotically tied.[1]

To this end, the Dutch introduced, in 1870, the Agrarian Land Law. Together with various ancillary enactments, this statute made it possible to transfer direct responsibility for insuring Java's profitability to private enterprises while preventing such enterprises from destroying the village economy upon which that profitability depended. In it, the convenient notion, current since the Raffles interregnum at the beginning of the

---

[1] It must be remembered, however, that although sugar production became steadily more capital intensive, it continued to use large quantities of unskilled seasonal labor on the cultivation side of its operations.

century, that all uncultivated "waste land" is inalienable state property was for the first time officially codified, making it possible for private plantation concerns to lease such land on a regularized, long-term contractual basis from Batavia and to use such legal titles for the purposes of obtaining credit. Further, the prohibition of outright alienation of peasant land to foreigners embodied in Javanese customary law was given formal legal backing by the government, permitting the systematization of rules and regulations for a similar long-term leasing of it to plantation companies for commercial use by its millions of small-holders. At base, the law, which inaugurates the Corporate Plantation period in East Indian economic history, represents one more effort to superimpose commercial economy upon subsistence economy in such a way as to stimulate the first and tranquilize the second.

The immediate beneficiaries of this legal innovation were the individual private Netherlands East Indies planters who had been created by the workings of the Culture System they professed to despise. This was particularly true in sugar, where privately planted cane made up about 9 percent of the total in 1870 and about 97 percent two decades later, though enterprise in coffee, tobacco, and even rice also appeared.[2] But the depression of the middle eighties (coffee prices fell by half, sugar prices by even more),[3] assisted by the onset of crop diseases in both coffee and sugar, wrecked the jerry-built financial structure of the Indies beyond repair. The self-styled "pioneers [and] speculative elements in the Indies itself," whom Reinsma celebrates, were consequently driven into the now waiting arms of large-scale Dutch finance, and they soon disappeared into enormous, multi-enterprise, limited-liability corporations rooted firmly in the motherland, or, in some cases, in other European countries. The Amsterdam and Rotterdam merchants of great traffic did not create, as they later came to claim, the Netherlands East

[2] Burger, 1939, p. 177.
[3] Furnivall, 1944, p. 196.

Indies estate economy. They bought it—and rather cheaply, considering the social costs of its production—at auction.

By 1900 the reorganization was well underway and the replacement of state mercantilism by corporate nearly complete. The Nederlandsche Handel Maatschappij (NHM), established as an imperial shipping company by William I in 1824 to prevent the English merchant marine from dominating the Far East and, when that effort failed, to carry Culture System products to Holland, transformed itself from crown's agent to private investment firm, half bank and half planter. By 1915 it owned, in whole or part, sixteen sugar factories, and had effective control over twenty-two others, plus fourteen tobacco, twelve tea and fourteen rubber estates.[4] By the same year, the Handels Vereeningen Amsterdam (HVA), founded as late as 1878, controlled fourteen sugar estates and managed one tapioca, one combined coffee-rubber and two fiber (sisal) plantations.[5] In the principalities, the dozens of independent planters of the Culture System period collapsed into the Cultuur Maatschappij der Vorstenlanden, which had twenty sugar enterprises by 1913, plus three in coffee, one in tobacco and one in tea. After the introduction of tobacco into northeast Sumatra in 1864, the Deli Maatschappij flourished and eventually spread its interests into rubber, oilpalm, sisal, tea and coffee.[6]

Backed by a network of banks—the Java Bank, the Nederlands-Indische Handelsbank, and others, as well as some foreign concerns, such as the Chartered Bank—the corporations diversified plantation production beyond the staple sugar and coffee of the nineteenth century, spread it to restricted parts of Outer Indonesia, built railroads and modern irrigation works, set up agricultural experimental stations to improve yields, and in general created a comprehensive agro-industrial structure probably unmatched for complexity, efficiency, and scale anywhere in the

---

[4] Encyclopedie van Nederlandsche-Indie, II, 53.

[5] Encyclopedie van Nederlandsche-Indie, II, 53.

[6] Encyclopedie van Nederlandsche-Indie, II, 53; Allen and Donnithorne, 1957, pp. 97–99.

world. By 1938 there were 2,400 estates in Indonesia, equally divided between Java and the Outer Islands, occupying about two and a half million hectares and controlled for the most part by a few large and, "under a coordinating superstructure of large syndicates and cartels," interlocked companies.[7]

But for all the structural changes within the Dutch side of dual economy which this transition to large-scale corporate activity brought about, it seems unlikely that the peasants noticed any essential difference, save perhaps that the pressure exerted on them was better organized. Indeed, even after the revolution, when the Dutch presence had just about evaporated, older Indonesians still tended to refer to the whole prewar complex of colonial institutions, political and economic alike, as "The Company." Nor, from their point of view, were they so much behind the times as it might appear. If ever the *plus ça change* maxim has held, it was with respect to the meeting of Dutch merchant and Javanese rice peasant in Indonesia.

Sugar, which remained Indonesia's most important export crop until the thirties, continued its mutualistic relationship with wet rice in but slightly changed guise. As cane was no longer grown on peasant land by government fiat and large-scale "waste-land" tracts suitable for it were scarce, now that Java was so crowded, a complex land-rent system was developed to obtain the use of sawah. A village, sometimes willingly, sometimes coerced by its leaders and local civil servants, contracted a 21½-year lease with an estate. The estate then planted one-third of the village sawah in cane. The cane occupied these fields for about fifteen months;

---

[7] Van de Koppel, 1946. Only about 1.2 million hectares or 47 percent of the land controlled by the estates was actually planted. A systematic, detailed history of the development of corporate enterprise in the Netherlands Indies remains to be written; but a general, rather routine, survey can be found in Allen and Donnithorne, 1957, esp. chapter x, pp. 181–199. For a literally bird's-eye view of corporation Java around 1927, plus a certain amount of unsystematized data on the various companies, see de Vries, n.d. The internal quotation is from Wertheim, 1956, p. 99.

after eighteen months the land was returned to the holders and another third of the village's land was taken for sugar, and so on around the cycle. But, as the new cane planting usually took place before the old one was harvested, any particular field was in sugar about a half rather than a third of the time; or to put it aggregatively, an average of about one-half the village's land, now one-third, now two-thirds, was in sugar, and half in peasant crops—either rice, or dry-season second crops such as soya or peanuts. One entire cycle therefore took three years, and seven such cycles could be completed during a single leasehold. The model example of such a cycle shown in Table 3 should clarify this only superficially complicated system, which in terms of formal aesthetics, if nothing else, must have appealed to the numerologically inclined Javanese.

The product of this complex and intimate interlocking of the rigorous cultivation rhythms of large-scale corporate agriculture and the more pliable ones of traditional household farming was an odd centauric social unit which was neither a proper plantation nor yet a peasant community, but different from either of them. The centaur's head was the mill. This heavily capitalized factory in the field (average outstanding investment in a single sugar enterprise seems to have run in the neighborhood of a million dollars during the thirties) with its steam- or electric-driven crushers, filters, centrifuges, evaporators, and vacuum cookers, was run by a European managerial staff of twenty men or so, settled with their families in neat bungalows huddled beside its walls, and processed the cane production of about a thousand hectares a year.[8] The centaur's body was the peasantry, upon whom the mill drew not just for land, but for the casual, seasonally extremely variable labor force it needed to clear fields,

---

[8] On sugar-milling technology, see Koningsberger, 1946. The investment estimate is based broadly on calculations with the very rough data given in Appendix II of Allen and Donnithorne, 1957, pp. 288–299, from which source the estimates of European staff and hectarage per mill also come (p. 84).

TABLE 3

A Typical Cropping Cycle for Sawah Land Leased
to Sugar Companies in Java after 1900

	1st third of sawahs	2d third of sawahs	3d third of sawahs
East monsoon (dry season) year 1	newly planted cane	harvestable cane	dry crops
West monsoon (wet season) year 1	growing cane	wet rice	wet rice
East monsoon year 2	harvestable cane	dry crops	newly planted cane
West monsoon year 2	wet rice	wet rice	growing cane
East monsoon year 3	dry crops	newly planted cane	harvestable cane
West monsoon year 3	wet rice	growing cane	wet rice
East monsoon year 4	the same as in the east monsoon in year 1		

Source: After Koningsberger, V. J., "De Europese Suikerrietcultuur en Suikerfabricatie," in van Hall, C. and C. van de Koppel, De Landbouw in De Indische Archipel, s'Gravenhage: van Hoeve, Part IIA, p. 326.

dig trenches, plant, harvest, carry cane to the mills, and perform countless other occasional tasks connected with the industry. By 1930, sugar concerns were employing more than 800,000 Javanese —men, women, and children—at one point or another during a year; or, as there were some 180 factories, an average of 4,000–

5,000 per mill.[9] Clearly, not all the "indirect-rule," "native-welfare," "East-is-East-and-West-is-West" colonial policies in the world could keep an encounter with big business of these dimensions from having a profound effect on rural life. And, as the leasing arrangement was also applied, in somewhat varying and less penetrating forms, to the commercial cultivation of other annuals—tobacco, cassava, agave, even rice—the effect spread, to some degree, over a fair part of the countryside.[10] As van der Kolff has said, in Java the clash of cultures came in the form of a clash of cultivations.[11]

Even more peculiarly distinctive of the Javanese situation—for the clash of cultivations has, after all, occurred elsewhere, and with perhaps even greater violence—is the fact that this clash took place largely within the context of a well-established, fully crystallized village social system, which though squeezed, deformed, and enervated was not destroyed by it. The sugar-lease system and, to a lesser extent, the similar practices connected with other crops, did not isolate the disequilibrating forces of commercial capitalism from village life; they introduced them, following the path the Company and the Culture System had blazed, into the very heart of it. Unlike, say, in Jamaica, the sugar industry in Java was not built up on the basis of imported slaves lacking peasant traditions. Unlike, say, in Puerto Rico, it did not forcibly marshal an embryonic peasantry onto enclave plantations and degrade it into a fully proletarianized, essentially landless labor force.[12] The Javanese cane worker remained a peasant at the same time that he became a coolie, persisted as a community-oriented household farmer at the same time that he became an industrial wage laborer. He had one foot in the rice terrace and the other in the mill. And, in order for him to main-

[9] Allen and Donnithorne, 1957; Boeke, 1953, p. 141.
[10] Boeke, 1953, p. 91.
[11] Van der Kolff, 1953.
[12] Mintz, 1951.

tain this precarious and uncomfortable stance, not only did the estate have to adapt to the village through the land-lease system and various other "native-protection" devices forced on it by an "ethical" colonial government, but, even more comprehensively, the village had to adapt to the estate.

The mode of its adaptation was again involutional. The basic pattern of village life was maintained, in some ways even strengthened, and the adjustment to the impingements of high capitalism effected through the complication of established institutions and practices. In land tenure, in crop regime, in work organization, and in the less directly economic aspects of social structure as well, the village, "hemmed in on all sides by a crystallized pattern" (to quote Goldenweiser again), faced the problems posed by a rising population, increased monetization, greater dependence on the market, mass labor organization, more intimate contact with bureaucratic government and the like, not by a dissolution of the traditional pattern into an individualistic "rural proletarian" anomie, nor yet by a metamorphosis of it into a modern commercial farming community. Rather, by means of "a special kind of virtuosity," "a sort of technical hairsplitting," it maintained the over-all outlines of that pattern while driving the elements of which it was composed to ever-higher degrees of ornate elaboration and Gothic intricacy. Unable either to stabilize the equilibrated wet-rice system it had autochthonously achieved before 1830, or yet to achieve a modern form on, say, the Japanese model, the twentieth-century lowland Javanese village—a great, sprawling community of desperately marginal agriculturalists, petty traders, and day laborers—can perhaps only be referred to, rather lamely, as "post-traditional."

The durability of the basic framework of the post-traditional rural economy is apparent no matter at which aspect of it one looks. On the side of land tenure, the so-called "communal ownership" systems under which the village as a corporate body exercises various kinds of residual rights of control over fields seem actually to have been strengthened, at least in relative terms, in the sugar-area villages.[13] The need on the mill's side for a simple, flexible, and comprehensive land-owning unit within which cane cultivation could move freely from one block of terraces to the next, unobstructed by a cloud of separate, individualized land rights, and the need on the villagers' side for a reasonably equitable sharing throughout the community of the burdens imposed by the system as it so moved from field to field, made the collective apportionment procedures of traditional communal tenure functional to both parties. With such devices, the mill could treat with the village as a whole through the agency of its headman and plan its regime in terms of ecological units large enough to be economic. And with the same devices, the village could insure—by means either of a periodic rotation and redivision of plots among qualified families or of a carefully adjusted scattering of the plots allotted such families through all three crop-sequence subareas—that at any point in the cycle none of its "landowners" would be entirely deprived of fields due to pre-emption by the mills. In brief, communal tenure allowed the mills to operate in terms of a large-field system appropriate to plantation sugar and, simultaneously, the village to operate in terms of the small-field system appropriate to sawah rice.

In this through-the-looking-glass world, even a major technical innovation in village agriculture—the emergence of unirrigated, annual food crops as true staples alongside paddy—acted more to dampen structural change in the rural economy than to strengthen it. With the exception of soya beans (and, of course,

---

[13] Van der Kolff, 1929; van Gelderen, 1929. Van der Kolff, in fact, traces the origins of this whole process, too, back into the Culture System period when, "if the indigo and sugar crops were to be a success they had to be grown in rotation on different sites, and since, from the point of view of supervision and irrigation, it was much easier to deal with compact blocks of land, it was to the advantage of the government to regulate matters with a powerful village council rather than with individual landowners" (van der Kolff, 1929, p. 111). For a general survey of traditional Javanese tenure systems, see Vollenhoven, 1906, pp. 604–634. For a more recent discussion, see van der Kroef, 1960.

unirrigated rice) all the dry-field annuals which are today important in Java—maize, cassava, sweet potatoes, peanuts—were introduced subsequent to European contact, and their penetration into the by then long-established sawah ecosystem was very gradual. In 1817, Raffles speaks of a recent increase in maize-growing on previously uncultivated poor-soil hill ranges "in the more populous parts of Java . . . where the sawahs do not afford a sufficient supply of rice," but other dry-field crops seem to be of but marginal importance.[14] Cassava began to spread over some less fertile areas in Bantam, Djapara, Semarang, and the Priangan after the experimental station in Buitenzorg imported shoots from Surinam in 1852, but it is only in this century that it has become an important diet item generally.[15] It was, in fact, not until the 1885 depression, when colonial officials actively promoted their use to increase native food production, that such crops—called collectively *palawidja*—began to play a vital role in the peasant economy.[16] By the first decade of this century they had become a fully integral part of village agriculture, both as east-monsoon second crops or sawahs and in crop-and-fallow dry fields, called *tegal*. Agricultural statistics before 1915 are scarce, contradictory, and unreliable, but Table 4 gives at least a global picture of the nature and magnitude of the change.

The necessity of using incommensurate time periods and variant classifications introduces a degree of uncertainty into the interpretation of this table beyond that which stems from the mere insubstantiality of the data upon which it is based. But that there was a burst of palawidja cultivation around the turn

---

[14] Raffles, 1830, I, 134–137.

[15] Koens, 1946. This was not the very first entry of cassava into the archipelago; Rumphius, "the blind seer of Ambon," reports it as early as the seventeenth century from the Moluccas.

[16] De Vries, quoted in Burger, 1939, p. 232. Scheltema (1930) quotes a *Koloniaal Verslag* for 1889 which claims that planted rice area increased about 8 percent over the 1817–1878 decade and that of palawidja by 24 percent, almost the whole (23 percent) of the latter increase coming in the form of second-crop planting on sawahs!

TABLE 4

THE GROWTH OF DRY-CROP CULTIVATION IN JAVA

	Estimated percent increase in amount of cultivated land in Java and Madura	
	sawah	tegal
1888–1900[1]	8	24
1900–1915[1]	16	150
1916–1928	10	26
1928–1938	3	5
	Estimated percent increase in harvested area	
	wet rice	palawidja[2]
1888–1904[1]	7	32[3]
1903–1920	15[4]	191
1916–1928	13	81
1928–1938	19	16
	Estimated percent of rice and nonrice crops in harvested area (peasant sector only)	
	rice[5]	nonrice
1888	65	35
1910	58	42
1920	51	49
1938	45	55

[1] Excludes the principalities and the *particuliere landerijen*. Were these included, the figures would probably be very slightly lower.

[2] Includes, in conflict with Javanese usage but in harmony with ecological reality, unirrigated rice (*gaga*). Relatively little dry rice is grown in Java in any case.

[3] 1888–1903.

[4] 1904–1920.

[5] Includes unirrigated rice, as breakdown not given in 1910 figures.

SOURCES: Scheltema, A. M. P. A., "De Sawahoccupatie op Java en Madoera in 1928 en 1888," Korte Mededeelingen van het Centraal Kantoor voor de Statistiek, Buitenzorg, 1930. Huender, W., Overzicht van den Economischen Toestand der Inheemsche Bevolking van Java en Madoera, s'Gravenhage: Martinus Nijhoff, 1921, pp. 35–37. Van Hall, C. J. J. Insulinde, De Inheemsche Landbouw, Deventer: Van Hoeve, n.d., pp. 216–217. Cabaton, A., Java, Sumatra, and the Other Islands of the Dutch East Indies, New York: Charles Scribner's Sons, 1911, p. 213. Statistical Pocketbook of Indonesia, 1957, Djakarta: Biro Pusat Statistik, pp. 47–48.

of the century is beyond doubt. The over-all pattern of development seems to have been one of a slow expansion of such planting, using nearby unirrigable land in a nonintensive, merely supplementary fashion, up to the end of the Culture System period; the penetration of it into the sawah complex proper in the form of the second-crop regime—ecologically, the decisive development—from about 1870 to 1900; and then a final, more intensive expansion of tegal clearing as virtually all of the remaining cultivable land of the island was brought under the plow. Sawah expanded too during this period, as a result of large-scale irrigation improvements, particularly along the northwest coast and in the Brantas-Solo delta; and the release by the sugar mills of more than three-quarters of their rented fields after the 1930 crash permitted a last-gasp expansion of the area cropped both in wet rice and, now that the dry monsoon regime was firmly established, in palawidja as well. But the really significant change is clearly the movement away from a typical Southeast Asian lowland rice monoculture toward a level of diversification unique for the region. By the mid-thirties the percentage of total peasant cropped land in rice in Java was, as the table indicates, about 45; in Thailand it was 90, in Indochina, 85, and in Burma and the Philippines, 65—that is, just about what it had been in Java in 1888.[17]

Yet, as noted, what is striking about this development is not how much it affected the general pattern of village economic life, but how little. Agricultural revolutions have been built on less, but in Java thorough-going diversification of traditional monoculture led only to the extension of the already far-advanced involutional process.

The penetration of maize, soyabeans, peanuts, and other crops into the sawah ecosystem did not change its essential structure. Poured into a rice mold, these crops merely reinforced it. They made it possible to drive a terrace just that much harder, to keep

---

[17] Wickizer and Bennett, 1941, p. 31; Thompson, 1946, p. 295; Scheltema, 1930.

pace with a rising population and, in the sugar areas, with the increasing external pre-emption of village lands, by increasing labor inputs into an already hyperintensive productive system. Thus, the all-Java ratio of harvested wet-rice acreage to total sawah acreage remained virtually constant between 1888 and 1928 at just about 1:1, increases in some regions being offset by decreases in others—mainly, as one would expect, sugar areas. But the ratio of harvested palawidja acreage *on sawahs* (i.e., terraces used during the west monsoon for paddy) to total sawah acreage rose about 25 percent from about .33:1 to .42:1, the main gains again coming, the special cases of arid Madura and infertile Rembang aside, in the principal sugar areas.[18] In short, the result of the expansion of palawidja cultivation was the same as that of virtually every other technological innovation in the peasant sector after about 1830—the maintenance of the marginal productivity of labor at some low but nearly constant (or perhaps gradually declining) figure. It merely gave the multiplying Javanese a bigger pool in which more of them could tread water.

This is no less true of the palawidja produced by crop-and-fallow farming on unirrigated tegals than of that grown on sawahs as east monsoon crops. Tegal cultivation did not lead to a significantly more extensive or more highly capitalized farming pattern nor, despite the freer entry of some of the crops into the market, to the growth of a commercial approach to agriculture. It led, rather, to the supplementation of intensive, small-scale rice-farming by intensive, small-scale dry crop-farming modeled almost exactly upon it. The size of the individual farm remained Lilliputian. Toward the rainy west it tended to be smaller with a greater percentage of sawah, toward the dry east larger but with a smaller sawah component. But it was essentially the same sort of farm—in fact essentially the same farm as had existed before the tegal expansion at the turn of the century. The ecological elasticity of wet rice having at last begun to fail him, the Javanese peasant turned toward diversified

---

[18] Calculated from Scheltema, 1930.

dry cropping to eke out, by the most labor-intensive pattern such crops could sustain, his customary meager living. And if one approaches the problem from the consumption side, one comes to the same conclusion as the Dutch agronomist Tergast:

On densely populated Java the rice production could not keep step [after 1900] with the increase in population. Increasing cultivation of other foodstuffs in rotation with rice and on dry lands has met the rising need for food. Around 1900 the amount of annual per capita quantity available was about 110 kg of rice, 30 kg of tubers, and 3 kg of pulses. Around 1940 this had changed to 85 kg rice, 40 kg maize, 180 kg tubers, and about 10 kg pulses. This change has locally reduced, often seriously, the quality of the diet. Expressed in calories there was little change between 1900 and 1940; the daily menu per capita has been kept on a level somewhat lower than 2,000 calories. That Java, in spite of the heavy population increase and the small possibility of expanding the cultivated area, has been in a position to maintain the calorie level of the diet is due principally to the intensification of rotation on rice fields and more intensive use of the dry lands.[19]

In addition to land tenure and land use, the involutional process also worked its peculiar pattern of changeless change on

---

[19] Tergast, 1950. Of course, Tergast's calculations are merely broad estimates also, perhaps even less precise than the more straightforward estimates of arable land and harvested area. In fact, it seems likely that there was actually a decline in caloric intake between 1900 and 1940. For a summary of other estimates of daily caloric intake (one of which sees it as declining from about 1,850 in 1921 to about 1,750 in 1938 and the other of which sees it as declining from about 2,000 to 1,900 over the same period), see van de Koppel, 1946. For yet another estimate of quantities of various food crops available per capita from 1913 to 1940 showing a pattern generally similar to Tergast's, see Pelzer, 1945, p. 259. In addition, both these calculations and my own analysis in the text omit, mainly because of measurement difficulties, consideration of the extreme intensification of grainless cultivation in house gardens (pekarangan), where on an average plot of some thirty square meters a Javanese family may grow literally dozens of different fruits, vegetables, and herbs, and produce, in extreme cases, as much as two-fifths of its caloric intake. See Terra, 1946.

the distribution side. With the steady growth of population came also the elaboration and extension of mechanisms through which agricultural product was spread, if not altogether evenly, at least relatively so, throughout the huge human horde which was obliged to subsist on it. Under the pressure of increasing numbers and limited resources Javanese village society did not bifurcate, as did that of so many other "underdeveloped" nations, into a group of large landlords and a group of oppressed near-serfs. Rather it maintained a comparatively high degree of social and economic homogeneity by dividing the economic pie into a steadily increasing number of minute pieces, a process to which I have referred elsewhere as "shared poverty." [20] Rather than haves and have-nots, there were, in the delicately muted vernacular of peasant life, only *tjukupans* and *kekurangans*—"just enoughs" and "not-quite enoughs."

By and large, the set of mechanisms producing this fractionization of output seems to have been centered less on land ownership than on land-working. Equal inheritance, reduction in size of communal shares to permit more shareholders in the same quantity of village land, and within-village bits-and-pieces land sale—all no doubt produced some diminution in the extent of individual holdings. But Javanese farms were small to start with. Though striking local exceptions can be found—again, particularly in the sugar areas where villagers were forced into more extreme measures—there is little evidence for any large-scale secular decline (or increase) in average farm size over Java as a whole during the colonial period. The estimates of mean—and modal—individual holding given by Raffles in 1817 and by Boeke in 1940 run just about the same: slightly under one hectare.[21] The greater part of the dispersion of output, which in

[20] Geertz, C., 1956.
[21] Raffles, 1830, I, 162; Boeke, 1953, pp. 52–143. Additional evidence that it was not, in the prewar period at least, changes on the ownership side which facilitated "sharing the poverty" comes from the fact that parcelization—the splitting of single holdings into a large number of very small

the nature of the case must have occurred, seems to have been mediated not through changes in the general structure of proprietary control, but through a marked elaboration and expansion of the traditional system of labor relations—in particular, of the institution of sharecropping.[22]

Javanese sharecropping consists of a large number of ad hoc variations on a handful of structural themes. It permits almost every case to be a special one, responsive to the peculiar circumstance of the individual contract, but at the same time a concrete example of a well-defined general type. It is this flexibility within fixity which has given it its peculiar usefulness as an involutional mechanism.

Against the background of a few precisely formalized, tra-

---

scattered plots—was not particularly extensive in Java (being about half the level of India or China), nor does it seem to have increased in the immediate prewar period. Further, leaving aside the "Outer Indonesia" region of Preanger only a (very grossly) estimated 2,000 Javanese owned more than 18 hectares of land in 1925. Scheltema, 1931, p. 275. One of the sources of the widespread notion of a radical decrease in average farm size in Java seems to be mere astonishment at its present smallness. Thus Mears (1961) writes: "With the increase in population density on Java, the size of individual holdings has declined to the point where approximately 90 percent of the farmers cultivate less than 1 hectare . . . of land. Almost 70 percent farm less than one-half hectare. Large holdings . . . are the exception with less than one percent of the agricultural holdings exceeding 10 hectares." When it is noted, however, that a survey taken in 1903, when the population was around half what it was in 1958 (and excluding the principalities where farms have always been if anything smaller than average), found that 15.8 percent of the farms were under .18 ha, 32.8 percent under .35 ha, 47.2 percent under .53 ha, 70.9 percent under .71 ha, 89.1 percent under 1.42 ha, and only 3.9 percent over 2.8 ha, the decline, though real, seems a great deal less precipitous and can in no sense account for the absorption of the bulk of the population increase into the village economy. Pelzer, 1945, p. 166.

[22] For an encyclopedic review of traditional forms of sharecropping in both Inner and Outer Indonesia, see Scheltema, 1931. For a more general discussion of the whole range of types of traditional agricultural labor relations, see de Bie, 1902, pp. 67–74.

ditionally rooted model procedures (e.g., the *maron,* or half-half, procedure) landowner and land tenant could reach a settlement concerning responsibilities for inputs (seed, oxen, labor, land taxes); could adjust the apportionment of output to whatever pre-existing social relationships obtained between them, as well as to type of land, type of crop, and so on; and could take into account the vagaries of local custom, individual moral sense, and momentary economic reality. Further, a nearly limitless divisibility of inputs. and outputs was possible through combining these procedures both with each other and with other sorts of labor agreements—subcontracting, renting, pawning, jobbing, work exchange, collective harvesting and, latterly, wage work.

The productive system of the post-traditional village developed, therefore, into a dense web of finely spun work rights and work responsibilities spread, like the reticulate veins of the hand, throughout the whole body of the village lands.[23] A man will let out part of his one hectare to a tenant—or to two or three —while at the same time seeking tenancies on the lands of other men, thus balancing his obligations to give work (to his relatives, to his dependents, or even to his close friends and neighbors) against his own subsistence requirements. A man will rent or pawn his land to another for a money payment and then serve as a tenant on that land himself, perhaps in turn letting subtenancies out to others. A man may agree, or be granted the opportunity, to perform the planting and weeding tasks for one-fifth of the harvest and job the actual work in turn to someone else, who may, in *his* turn, employ wage laborers or enter into an exchange relationship with neighbors to obtain the necessary labor. The structure of land ownership is thus only an indifferent guide to the social pattern of agricultural exploitation, the specific form of which emerges only in the intricate institutional fretwork through which land and labor are

---

[23] In fact, the "share-crop" principle is applied not only to agriculture but also to farm animals. See Scheltema, 1931, pp. 243–252.

actually brought together.[24] In share tenancy and associated practices the ever-driven wet-rice village found the means by which to divide its growing economic pie into a greater number of traditionally fixed pieces and so to hold an enormous population on the land at a comparatively very homogeneous, if grim, level of living. What elsewhere has been sought through land reform—the minimization of socioeconomic contrasts based on differential control of agricultural resources—the Javanese peasant, whose farms were minute to start with, achieved through that more ancient weapon of the poor: work spreading.

On the technical level this choice, if one can call it that, to diffuse increases in per-hectare productivity through a larger labor force, thus maintaining a constant or near-constant per-worker productivity, has the effect of leading toward more fixed (or presumed fixed) coefficients of production on the peasant side similar to (but precisely reversed from) those on the plantation—or, more accurately, to coefficients variable only in one direction. Once the radical intensification of agriculture is accomplished, it is difficult to retreat from it. Sawah terraces are pushed up the sides of mountains and along the faces of river banks, where their narrowed form often means a reversion to plowless cultivation. Land fragmentation, so far as it occurs, has a similar effect, because as the edge of the terrace must always be worked with a hoe, the smaller the terrace the less advantageous the use of the plow. (Even in the absence of ownership fragmentation, florescence of in-kind tenancy patterns would seem to stimulate a reduction in the size of individual terraces, as would more labor-intensive methods generally, though this is impossible to prove.) Double cropping, more careful micro-regulation

---

[24] Aside from often applying to the same man at the same time, the terms "landlord" and "share tenant" may even be slightly misleading in their usual connotation in this context where, as Boeke remarks, the cropper is often the stronger party. 1953, p. 58. See also, in this connection, Burger, 1930.

of irrigation in and around the terrace, stalk-by-stalk, not to say grain-by-grain, cultivation, reaping and milling methods, developed share-cropping, and other labor innovations all are difficult to abandon once they are firmly instituted. Thus, as the Dutch planter comes to feel that the only possible direction of increased rationalization is in the direction of greater investment in machinery, modern irrigation, scientific experimentation, and so on, the Javanese peasant comes to feel that the only feasible way at least to maintain his living standard, much less raise it, is through more fine-comb agricultural methods, and tends to become increasingly skeptical of the possibilities of improvement through mechanization. Where the plantation sector becomes addicted to capital, the peasant sector becomes addicted to labor —the more they use the more they need. A sort of mirror-image ratchet effect, to change the metaphor, begins in which it becomes relatively easy for the planter to increase capital inputs, but difficult to reduce them; and relatively easy for the peasant to increase labor inputs but difficult to reduce them. The productive process of the one sector becomes steadily more capital-intensive and that of the other steadily more labor-intensive, until the dualistic gulf which yawns between them seems unbridgeable, as it did to Boeke.

The conservation of traditional land-tenure systems, the assimilation of dry crops to the wet-rice pattern of land use, and the elaboration of established systems of labor relationship are thus all of a piece. They represent a reaction—for all its oft-admired ingeniousness, almost wholly defensive—to the twin pressures of a rising population and a superimposed plantation economy. But they do not exhaust that reaction, for it was as much a cultural and social structural process as an economic and ecological one. The involution of the productive process in Javanese agriculture was matched and supported by a similar involution in rural family life, social stratification, political organization, religious practice, as well as in the "folk-culture" value system—

what I have elsewhere called the *abangan* world view—in terms of which it was normatively regulated and ethically justified.[25]

Thus, at each stage of the development which we have been following on the level of the rice terrace and the cane field, we could presumably have traced similar processes in the various social and cultural institutions which comprise the backbone of village life. The elaboration of a complex ranking system tied to communal tenure, the fluctuating relationships between kinship, patron-dependent and territorial ties, the vicissitudes of the position of the village headman and his staff, the changing content of village rights as a corporate unit, and even the transformations of the *abangan* "little tradition" as it absorbed into itself a wide range of foreign elements and turned them to its own ends—all these would add substance to the "cultural core" side of our description of the history of a human ecosystem. But if statistics are uncertain before this century, community-level ethnographies are nonexistent; and even the handful of village studies carried out in the closing phases of colonial rule are, though excellent in themselves, almost wholly oriented to economic and agrarian matters—to an obsessive concern with that elusive entity, "native welfare"—rather than to the close analysis of social structure and culture patterns.[26] By looking at the range of variation among contemporary villages, by squeezing dry what few fragments of local ethnography we have from the colonial period, and by deductively employing some general sociological principles, we can get an over-all sense for the things which must have happened to the Javanese village in its journey toward post-traditionality, but the details of "what happened" are inaccessible to us.

Perhaps the most trenchant phrase which has been coined to summarize what seems to have been the career of the Inner Indonesian village over the past century and a half is "the ad-

[25] Geertz, C. 1960.

[26] For some of the more important such studies, see Adam, n.d.; Burger, 1928; Burger, 1930; van Doorn, 1926; and van der Kolff, 1937.

vance toward vagueness." The peculiarly passive social-change experience which, especially in "inner" Inner Indonesia, rural society has been obliged to endure seems to have induced in it an indeterminateness which did not so much transform traditional patterns as elasticize them. Such flaccid indeterminateness is highly functional to a society which is allowed to evade, adjust, absorb, and adapt but not really allowed to change, much as similar traits are useful on an individual level to persons contained in "total institutions"—prisons, mental hospitals, concentration camps, and the like.[27] Pulled this way and that, hammered by forces over which it had no control and denied the means for actively reconstructing itself, the village both clung to the husks of selected established institutions and limbered them internally in such a way as to permit a greater flexibility, a freer play of social relationships within a generally stereotyped framework. The result was an arabesque pattern of life, both reduced and elaborated, both enormously complicated and marvelously simple: complicated in the diversity, variability, fragility, fluidity, shallowness, and unreliability of interpersonal ties: simple in the meager institutional resources by which such ties were organized. The quality of everyday existence in a fully involuted Javanese village is comparable to that in the other formless human community, the American suburb: a richness of social surfaces and a monotonous poverty of social substance.[28]

The Development of Outer Indonesia

But while Java was attaining the higher reaches of involution, a different process was taking place in the Outer Islands— or, more accurately, in some narrowly delimited sections of them. There, the enclave, not the mutualistic, pattern was flourishing, a true proletariat was being forged from a mass of imported

[27] Goffman, 1961.
[28] For discussions of some "suburban" aspects of contemporary Javanese villages, see Jay, 1956; and Geertz, 1959. I have incorporated a few phrases from the latter paper in the paragraph above.

coolies, and small-holder cultivation of export crops was increasing at an accelerating pace. The simplified statistics of Table 5 present the outlines of a pattern of dualistic development much more comparable to the segregative one of Malaya than to the symbiotic one of Java.

TABLE 5

VALUE OF NINE LEADING EXPORTS, JAVA VERSUS THE
OUTER ISLANDS, 1870-1930

	Indexes		Percentage of N.E.I. exports	
	Java	Outer Islands	Java	Outer Islands
1870	100	100	87	13
1900	172	400	70	30
1930[1]	555	3671	44	56

[1] All petroleum accounted to Outer Islands.
SOURCE: Calculated from J. S. Furnivall, Netherlands India, New York: Macmillan, 1944, p. 337.

Three characteristics of this Outer Island (or Outer Indonesia) development, each of them sharply contrasting to the Javanese pattern, serve to define it.[29] First, the development was geographically extraordinarily localized, particularly on the capital intensive side. Of the 1930 Outer Island export value, 600 million guilders, about a third originated from the East Coast of Sumatra alone, where the great tobacco, rubber, tea, and oilpalm plantations were concentrated in an area (Deli and its environs) downward of 10,000 square kilometers, or less than one percent of the total Outer Island area. Another third came from the oil fields

---

[29] If the figures underlying Table 5 could be broken down along the Outer versus Inner Indonesia divide, thus throwing the west Javanese tea development on the other side of the line where most of it properly belongs, the picture given would be even more dramatic.

clustered around Palembang-Djambi in south Sumatra and Balikpapan in east Borneo. And the tenth represented by tin came entirely from the three small islands of Bangka, Bilitung, and Singkep off the Sumatran south coast.[30] Second, rather than being focused on condiments, confections, and stimulants the development centered on the production of industrial raw materials—a reflex, as Wertheim has pointed out, of the alteration in world market conditions attendant upon the spectacular growth of large-scale manufacturing in the West after the middle of the nineteenth century.[31] In 1900, rubber, tin, and petroleum —almost entirely Outer Island products—accounted for about 17 percent of Indonesia's export value; in 1920 about 20 percent; in 1930 about 37 percent; in 1940, after the collapse of sugar in the depression, about 66 percent.[32] And third, as indicated earlier, the peasantry played a relatively much greater role in the export economy, the small-holder share of the 1930 Outer Island figure being about 35 percent (all of it, of course, from agricultural rather than mineral production) as against a comparable Javanese figure of about 15 percent.[33]

Except for tin, which was mined on Bangka as early as the seventeenth century, virtually the entire Outer Island development can be dated from the impulse in 1863 of the obscure tobacco planter Jacob Nienhuys to move from East Java to Deli.[34]

---

[30] Furnivall, 1944, p. 336; de Ridder, n.d., p. 77; van de Koppel, 1946, p. 393. At the end of the twenties, three-quarters of the total planted estate area outside of Java was situated in the residency of the east coast of Sumatra (van Gelderen, 1929).

[31] Wertheim, 1956, pp. 67, 97.

[32] Furnivall, 1944, pp. 36–37; Statistical Pocketbook of Indonesia, p. 135.

[33] Furnivall, 1944, p. 337; van de Koppel, 1946, p. 411. Taking agricultural exports alone (i.e., excluding tin and petroleum), the Outer Island small-holder share was about 53 percent as against the Javanese 17 percent. Van de Koppel, 1946. In per-capita terms the contrast is, of course, more striking: in 1927, about 21 guilders per head in the Outer Islands, about 2 in Java (van Gelderen, 1929); in 1939, 14 and 2 (Boeke, 1947, p. 25).

[34] On Nienhuys, see Boeke, 1953, pp. 220–221, 223; and van der Laan, 1946.

In a Surabaja trading house, Nienhuys, a Schumpeterian entrepreneur, met a certain Abdullah, an Arab who, passing himself
off as a Sumatran prince, spoke glowingly of the native-grown
tobacco of the island's little-known northeast coast and of the
profits to be made from it. He may not have been a prince, but
he was an Arab, and his sense for commercial possibility unerr-
ing. Having failed in Java, Nienhuys gathered together a handful of doubtful credits and departed forthwith to Deli to found
what became in time one of the most profitable tobacco enterprises—per unit of tobacco produced, *the* most profitable—in
the world.[35]

From the ecological point of view, tobacco was in many ways
the ideal crop to serve as bridge between the sawahs of Inner
Indonesia and the swiddens of Outer. An unirrigated, commercial annual with a short growing season and no technical dis-
continuities in scale, it could be fitted into virtually every sort of
agricultural pattern found in Indonesia. In Java it was grown as
a palawidja small-holder crop on dry-monsoon sawahs and on
unirrigated tegals, the only nonfood crop so cultivated in significant quantities.[36] In Sumatra, and sporadically elsewhere in
the Outer Islands, it was integrated into the multicrop fabric of
the swidden system, the pattern Nienhuys found established in
Deli.[37] On the estate side the same gamut appears. In the central
Javanese principalities (Jogjakarta and Surakarta) a version of
the rotating lease system already described for sugar was employed, with the estate conscripting the peasants' land and labor
every other year for tobacco cultivation under European direction.[38] In the other important estate tobacco area of Java, the

---

[35] In 1938 the world price of Deli tobacco was four to five times higher
than that of any other cigar tobacco, including Cuban. Van der Laan, 1946.
At the beginning of the century some Deli tobacco companies were paying
dividends as high as 111 percent. Encyclopaedia Britannica, 1911.

[36] Van den Broek, 1946.

[37] Van den Broek, 1946; van Hall, n.d., p. 69; Lekkerkerker, 1916, p. 243.

[38] Beets, 1946. As the leasing system in the principalities was modeled on
the so-called appanage system by which the royal courts of the indigenous

East Hook, the peasants grew the crop themselves and merely sold it to the estates, though by providing seedlings, paying land taxes, and advancing premiums the estates soon contractually bound them hand and foot to do so.[39] And, finally, in Sumatra yet another ingenious leasing system was devised, which permitted the Dutch planter to become a swidden farmer, like the local peasant.

. . . the famous Deli wrapper tobacco is grown only on land which has been lying under forest fallow for at least seven or eight years. . . . The tobacco planter of East Sumatra is thus a shifting cultivator who raises tobacco only once in a cycle of eight years, and each tobacco plantation therefore requires about eight times as much land as is planted in any one year.

. . . When the tobacco companies originally applied to the sultans for agricultural concessions, no attempt was made to set aside sufficient land for the peasantry, although the contracts stipulated that the companies had to give each peasant family 4 *bouw* (2.9 hectares) of land so that it could carry on its own traditional form of agriculture. But instead of reducing the size of their concessions by turning over to each family within the concession the stipulated amount of land, the planters preferred to have the use of all the land and to lend the harvested tobacco land to the peasants raising a single rice crop. This meant that the planter had access to all but the actual village land within his concession, while the East-Coast peasant no longer had to clear land for his own [swidden]. Instead, he received from the estate each year in May or June the same amount of land ready for the sowing of upland rice that he would have had to clear had he not lived within a tobacco concession.[40]

---

state of Mataram were supported, the peasant rendering three-fifths of his crop to the king or his agents, the provision of both land and labor to the estates was compulsory. Also as tobacco is only a dry-season crop, during the year the estate had possession of the peasant's sawah a wet-rice crop could also be grown (called, with fine irony, "permitted paddy") which was share-cropped on the usual half-and-half basis, with the estate as the landlord and a peasant, commonly the land owner, as the tenant.

[39] Boeke, 1953, pp. 212–213.

[40] Pelzer, 1957. Actually, the eight-year cycle was not so rigid; in 1940,

At first glance, it might seem that this relationship between swidden tobacco and swidden rice was also symbiotic; and Pelzer uses that term to describe it.[41] But in ecological, or at least cultural-ecological, terms it was not symbiotic, if by that is meant a relationship from which one or both ecosystems benefit and neither is harmed.[42] Rather it was neutral (i.e., a relationship which has no significant effect on either system) or, perhaps, mildly antagonistic (i.e., one which affects either or both adversely).[43] In Java, the sugar estate and the rice village came to occupy complementary niches in one, closely integrated wider

about 7 percent of the tobacco land was recultivated in seven or less years, about 33 percent in nine or more years. Van der Laan, 1946.

[41] Pelzer, 1957.

[42] Clarke, 1954, pp. 363–364. Clarke's conceptualization is cast in terms of species but is clearly extendable to ecological populations, communities, systems, etc.

[43] Clarke, 1954, pp. 363–364. By "beneficial" and "harmful" are meant, of course, the implications of the relationship for the equilibrium and persistence of the system qua system, not its value in terms of human welfare, so that "eufunctional" and "disfunctional" might be better terms (see, Levy, 1952, pp. 76–82). Using a plus to indicate a eufunctional ("beneficial") association, a minus to indicate a disfunctional ("harmful") one, and zero to indicate a functional neutral one, Clarke presents the following paradigm:

Species [ecosystem] A	Species [ecosystem] B	Relation	
plus plus	plus zero	mutualism commensalism	symbiosis
zero	zero	toleration	neutrality
zero plus minus	minus minus minus	antibiosis exploitation competition	antagonism

In terms of these rubrics, which of course shade into one another, the sugar-sawah system falls under mutualism, the tobacco-swidden system under toleration or (mild) antibiosis.

ecosystem; they were mutually adapted to one another in such a way that each made a positive contribution to the other's viability. In Deli, the tobacco estates and the autochthonous peasantry operated in parallels; they were not integrated into a single ecosystem but into two separate ones of a somewhat similar type. They were functional alternatives, not complements, and their joint persistence was the result not of positive mutual adaptation but of an elaborate program for keeping out of one another's way.

It was, of course, the swidden cultivator who for the most part was obliged to keep out of the way of the tobacco planter, rather than the other way around. There were few of him in any case: in 1915, the autochthonous population density in east Sumatra seems to have been less than six per square kilometer, which is low even for a swidden area.[44] Further, the evolution of the legal framework governing the concession contracts—which were, at least in the formative period of the industry, made between the local harbor sultans and the tobacco corporations, not between peasants and the corporations—steadily constricted the scope of operation for swidden cultivators. In 1884, when the rights of the peasants within concession areas were first given explicit legal form, the contracts guaranteed "a harvest year" to the peasant following tobacco cultivation, but by 1892 they guaranteed only "a harvest." Where the contracts originally had permitted the growing of "rice *and* maize," they now permitted only "rice *or* maize." Where cultivation rights had been allowed to anyone holding traditional, customary-law tenure claims over land in the new concession, only those actually *dwelling* on the concession—often a relative matter for a swidden farmer—were

---

[44] Calculated from Lekkerkerker, 1916, pp. 271, 339. In calculating the size of the indigenous population I have subtracted only the total of extra-Sumatran Indonesian coolies from the "natives" total, ignoring any families they may have had and whatever Indonesians, Sumatrans, or "foreigners" were engaged in noncoolie occupations, so that in fact the estimate of the swidden peasant density is a maximum, the density probably being in fact closer to four per square kilometer than six.

to be granted such rights. Where all harvested tobacco fields had been accessible, only those which peasants had in fact earlier cultivated were considered as open. And where the area allotted to each family had been a fixed quantity, the total land to be given out for swidden was henceforth limited to no more than half the cleared tobacco fields, thereby preventing any significant increase in the number of peasant cultivators on the concession.[45] Thus, the swidden cultivator had every reason to suspect that the tobacco planter really did not want him around at all. In fact, when the system of land concessions granted by local lordlings was abandoned after 1919 for a straightforward "waste land" type of lease from the Netherlands East Indies government —one legally free of encumbrances by customary native rights— the estates began to consider simply relinquishing a portion of their concessions to the swidden population outright in exchange for being permitted to bar native cultivators from tobacco lands altogether—a complete divorce which was about to be finally consummated when the colonial period ended.[46]

The separation between tobacco plantation and swidden farm is even clearer on the labor side. In contrast to Java, the local population of east Sumatra was not drawn upon for estate work, both because it was quantitatively insufficient and, having enough land and freedom, psychologically disinclined. Instead, first Chinese and later Javanese workers were brought in as a fully proletarianized indentured labor force. In 1913 about 48,000 Chinese and 27,000 Javanese coolies worked on east Sumatra tobacco estates, and of the total work force of more than 85,000, less than 1,000 were natives of the area itself.[47] Brought there by contract and held there by the notorious "penal sanction" which

---

[45] De Ridder, n.d., p. 45.

[46] Van der Laan, 1946. For a review of the variety of legal forms under which Western enterprises gained title to Indonesian land, see Allen and Donnithorne, 1957, pp. 67–70; and van de Koppel, 1946.

[47] Lekkerkerker, 1916, p. 272. More than 95 percent of the over-all total were contract coolies but of the small local group 95 percent were free. (By 1931, however, almost half of all coolies, now some 350,000, were free. Furnivall, 1944, p. 356.)

provided prison terms for runaways, this uprooted labor force was as much an enclave in indigenous Deli society as was the Dutch managerial class. The Malays and Bataks of Deli were in the European estates but, unlike their counterparts in Java, not of them.

The swidden ecosystem was, therefore, not seriously altered, even in an involutional manner, by its encounter with the estate tobacco system. In the first place, unlike the sawah system, it was incapable of such alteration; any serious attempt to intensify it would have led to grass-field erosion, and would have harmed both systems rather than benefit them in the way that improved irrigation, better land care, and crop diversification benefited them in Java. In the second place, the Sumatra estates, facing a labor shortage and a land surplus, rather than the reverse, did not see the indigenous population as a usable resource in the way the sugar planter saw the population of crowded Java, but merely as a local nuisance, like mosquitoes.[48] Whatever effects the estate regime had on swidden were probably—there are, to my knowledge, no detailed studies on the matter—in the direction of impoverishment. The fact that the swidden farmer did not have to clear the land of forest was, of course, an important labor-saving for him; but that he was restricted to planting second-season land, largely confined to a single crop, and limited as to the area he could plant, must have reduced the per-capita productivity of the regime. The over-all result was, in all likelihood (though, again, specific data to prove it seem to be unavailable), to thin further the peasant population living on the estates, or at least to lead those who lived on them to clear land outside of them as well.

In any case, as the Deli economy diversified beyond tobacco to shrub and tree-crop perennials—oil palm, agava, tea, and especially rubber—the ecological insulation of the peasant from the estate became virtually absolute, because no swidden farmers lived on such estates (see Map 6). The movement to diversifica-

---

[48] For a similar argument, a little differently put, see van Gelderen, 1929, esp., p. 100.

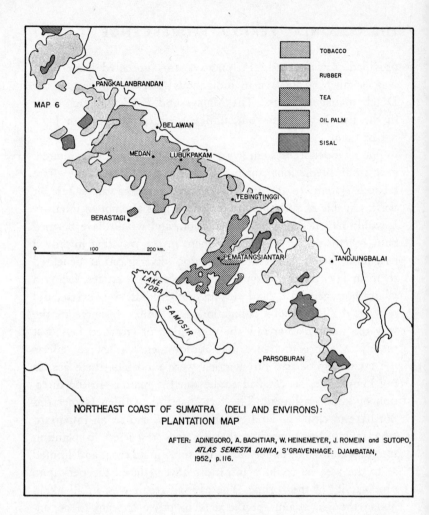

	TOBACCO
	RUBBER
	TEA
	OIL PALM
	SISAL

MAP 6

PANGKALANBRANDAN

BELAWAN

MEDAN   LUBUKPAKAM

TEBINGTINGGI

BERASTAGI

0   100   200 km.

PEMATANGSIANTAR   TANDJUNGBALAI

LAKE TOBA

SAMOSIR

PARSOBURAN

NORTHEAST COAST OF SUMATRA (DELI AND ENVIRONS):
PLANTATION MAP

AFTER: ADINEGORO, A. BACHTIAR, W. HEINEMEYER, J. ROMEIN and SUTOPO,
ATLAS SEMESTA DUNIA, S'GRAVENHAGE: DJAMBATAN,
1952, p. 116.

tion began early, for even the hardy Nienhuys tried to grow black
pepper, nutmeg, rice, opium, and coffee as well as tobacco on his
newly found land.[49] But it was not until the rubber industry
took hold that the whole direction of the Deli complex, and in
fact of the whole Outer Indonesian economy, changed drastically.
Planted first (in the form of *ficus elastica*, the Asian species) on a
very small scale in Java in the 1860's, rubber was not introduced
into Sumatra's east coast until 1906, long after tobacco was well

[49] Boeke, 1953, p. 223.

established, and subsequent to the famous abduction of *havea brasiliensis* to Malaya via the Kew Gardens. But once there, it initiated "one of the most remarkable periods of development in the history of agriculture." [50] Between 1913, when the trees first began to become tappable, and 1922, Outer Island estate-rubber production (about three-quarters of which derived from Sumatra's east coast) went from 3,000 metric tons to 43,000; in another decade (1931) it had reached 95,000; by the end of the colonial period (1940) 182,000—approximately one-third of the total world production. [51]

It was with respect to rubber that the small-holder export economy of the Outer Islands really "took off." Tobacco was effectively barred from most of the tropical-forest environment, except for small patches in the swidden fabric. It flourished in Deli because of the hyperextensive nature of the regime and because the northeast coast is one of the very few areas of Sumatra enriched by recent volcanic action. [52] But rubber is a tropical-forest product *par excellence,* a tree crop as readily integrated into a swidden complex as sugar was into a sawah complex, and hence spread rapidly and easily from its east-coast beachhead through most of Sumatra, to parts of Borneo and, to a lesser extent, to the Outer Indonesia sections of Java (Bantam, Priangan, Buitenzorg, and Batavia); by 1938 an estimated 60 percent of Netherland East India rubber exports by weight came from

---

[50] Allen and Donnithorne, 1957, pp. 117–118.

[51] Robequain, 1954, p. 166; Maas and Bokma, 1946; van de Koppel, 1946. Estate production rose on Java too, where rubber was sometimes "mixed" with coffee or tea on a single plantation in a lowland-highland arrangement. The comparable Java figures are 1913, 1,000; 1922, 29,000; 1940, 99,000 metric tons. Maas and Bokma, 1946. Unlike almost all other estate cultivations which were predominantly Dutch, somewhat less than half the investment in rubber was Dutch, and the largest single estate was American. A somewhat similar pattern, a little differently organized, occurred also in petroleum, where nearly 60 percent of the investment came from the Anglo-Dutch "Shell" combine and about a third from the United States. Allen and Donnithorne, 1957, pp. 288–289.

[52] Mohr, 1945.

nearly 800,000 small holders.[53] The colonial paradox held once again: ecological separation reduced rather than accentuated economic contrast. The Indonesians whose relationship to the estates was indirect benefited most from them.

Stimulated by the demonstration effect of rubber and paced by the general upsurge of commercial spirit in the Outer Islands after the turn of the century, the small-holder cultivation of other export perennials also flourished, at least until the 1930 depression. Copra export volume, nearly all of which came from small holders, grew more than sevenfold between 1900 and 1930.[54] Coffee exports from Outer Island small holders almost doubled, both by weight and value, in three short years (1925–1928) coming thereby to account for about half the total coffee production and nearly three-quarters of the coffee export of the Netherlands East Indies generally.[55] Pepper, exclusively a swidden small-holder crop since the days of the Company and before, spurted also, exports going from 25,000 to 40,000 tons between 1910 and 1925.[56] And even tea, until about 1910 almost exclusively an estate crop, was increasingly cultivated by small-holders, especially in the Outer Indonesia parts of west Java, reaching nearly a quarter of the marketed product by 1926.[57] But in Inner Indonesia the peasant economy continued to ex-

---

[53] Van Gelder, 1946. Because of the generally inferior quality of small-holder rubber as against estate rubber, the percentage by value was somewhat lower; about 45 percent (1937). Boeke, 1953, pp. 128–129; Allen and Donnithorne, 1957, p. 291. Between 1925 and 1940 small-holder rubber area increased from about 300,000 ha. to about 1,220,000 ha. Thomas, n.d., p. 21.

[54] Reyne, 1946; van de Koppel, 1946.

[55] Paerals, 1946; Boeke, 1947, pp. 23, 68; Allen and Donnithorne, 1957, pp. 291–292. About 9 percent of Netherlands East Indies small-holder coffee was produced on Java in both years, so that it also increased in absolute terms; virtually all of it was domestically consumed, however. Paerals, 1946.

[56] Rutgers, 1946. In this case, the advance continued, even accelerated, after the depression, reaching nearly 80,000 tons in 1935.

[57] Van Hall, 1946.

pand only statically. Squeezed by paddy, small-holder sugar never got off the ground; squeezed by palawidja food crops, small-holder tobacco, though much more significant than sugar, seems actually to have lost ground to estate cultivation over the first third of the century—at least as an export product.[58]

For the general pattern of peasant life, the outcome of this radical social change—which raised the peasant share of total Netherlands East Indies export value from about 10 percent in 1894 to nearly 50 percent in 1937—was different from the involutional pattern traced for Java.[59] But if prerevolutionary Javanese community studies are scarce, none exist for the Outer Islands, so that it is impossible to trace the transformations wrought in swidden and in the cultural "core" within which it is set by this commercial revolution. Looking here and there and adding bits and pieces, we can, at best, get an over-all picture whose outlines may be bold but whose details are dim.

The first such outline is that the impact of small-holder commercialization on the huge Outer Indonesia mass was uneven and highly focused. More than a quarter of the small-holder rubber trees were (1938) in the south Sumatra residency of Palembang alone; if the east coast of Sumatra, Djambi (the Sumatra central coast), and the western and southeastern corners of Borneo are added, nearly 80 percent are accounted for.[60] Of the total copra export, more than a third came (1939) from the Minahasa residency in northern Celebes; if the western edge of Borneo and the tiny Riau archipelago are added, the total

---

[58] Van de Koppel, 1946; van den Broek, 1946. Most Javanese small-holder tobacco production went either into "direct consumption" (i.e., hand-rolled straw-wrapped cigarettes), the tobacco for which was sold in shredded bulk form in local markets, or into domestic cigarette industries. In 1935, of a total Javanese production of 66,000 tons, about 55 percent was "directly consumed," about 25 percent worked by domestic (Chinese and Javanese) cigarette industries and only about 20 percent was exported. Van den Broek, 1946.

[59] Van Hall, n.d., p. 205.

[60] Van Gelder, 1946.

rises to three-fifths.[61] Eighty-five percent of the pepper came (1935) from the Lampong districts at the southern tip of Sumatra, 95 percent of the peasant-grown tea from the Priangan highlands of west Java.[62] Small-holder coffee was less focused, but even there the bulk of the output came from central and south central Sumatra, with north Bali, central Celebes, and Timor as secondary centers.[63] The growing points of peasant enterprise in Outer Indonesia were themselves islands in a broad sea of essentially unchanged swidden-making. As the Inner Indonesian heartland filled up to become a largely homogeneous, post-traditional rural slum, the Outer Indonesian arc differentiated into a plurality of sharply distinct enclaves of social and economic dynamism scattered through a monotonous expanse of enduring stability.

One such enclave on which (as a result of a single study done by the brilliant Dutch sociologist Schrieke in 1928) some circumstantial material exists concerning socio-economic change in Outer Indonesia as a result of the spurt in small-holder export cultivation, is the Minangkabau region of west central Sumatra.[64] This region—roughly the present residency of west Sumatra (Sumatra Barat)—is not the most typical example one might have chosen to represent the Outer Indonesia development generally. Not only is it the most densely populated area outside of Java and Bali (1930), but also one of the few places in the Outer Islands where irrigated rice was (and is) to be found on a fairly extensive scale, and perhaps the only one where small-holder coffee, rubber, and copra all came to be widely cultivated.[65] Yet, at the same time, this ecological heterogeneity,

---

[61] Reyne, 1946.

[62] Rutgers, 1946, van Hall, 1946.

[63] Paerals, 1946.

[64] Schrieke, 1955, pp. 83–166. Schrieke's study (The Causes and Effects of Communism on the West Coast of Sumatra) was originally presented as the summary report of an Netherlands East Indies Government Investigation Committee set up, with Schrieke as chairman, to look into Communist-inspired political disturbances on the west coast in 1926.

[65] Indonesia, 1956, I, 81; Schrieke, 1955. The Minangkabaus are the

combining sawah and swidden elements, makes the Minangkabau case of the commercial transformation of shifting cultivation all the more telling. Because Minangkabau agriculture developed differently in response to external challenges and pressures it is a particularly instructive example.

The Minangkabau cultural hearth lies in the Agam plateau, a nine-hundred-meter-high tuff tableland strangely corrugated by a series of broad, flat-bottom canyons, which, winding among a cluster of irregular schistose hills and towering volcanic cones, joins the three classical capitals of Bukit Tinggi, Pajakumbuh, and Batu Sangkar (see Map 7).[66] In this landscape the sawah system, aided by effusions from the cones, crystallized, and the character of Minangkabau culture—an unusual fusion of Islamism, matrilineality, and gnomic moralism—took form.[67] But around this hearth—in the narrow coastlands to the west (Pariaman, Padang, Painan), the flat marshes to the east (Bangkinang, Taluk, Sungaidareh), the highland river galleries to the north (Libuk Sikaping, Panti, Rau), and the rugged Kerintji mountain areas of the south (Muara Labuh, Sungai Penuh)—lies an enveloping tropical forest frontier into which the Minangkabau and their culture have evidently been spreading for centuries. It is in this periphery that swidden was concentrated, that peasants encountered at least the distant shock waves of the estate explosion, and that Schrieke found "the beginnings of an agrarian revolution":

---

most numerous people in Sumatra, forming about one-quarter of its population. Robequain, 1954, p. 157.

[66] Lekkerkerker, 1916, pp. 32–35; Robequain, 1954, pp. 157–158. Agam (or "old Agam") is perhaps more often used to refer to the plateau directly around Bukit Tinggi, different names being given to the linked ones of Pajakumbuh and Batu Sangkar.

[67] Robequain, 1954. A secondary sawah development is to be found on the lower (350 meters) Solok plain directly south of Lake Singkarak. Minangkabau sawah technique is different from Javanese, being in general less labor intensive. See van der Veer, 1946; and Terra, 1958.

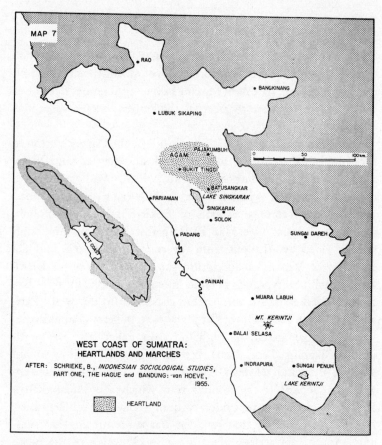

MAP 7

RAO

BANGKINANG

LUBUK SIKAPING

AGAM
PAJAKUMBUH
BUKIT TINGGI
BATUSANGKAR
PARIAMAN
LAKE SINGKARAK
SINGKARAK
SOLOK
WEST COAST
PADANG
SUNGAI DAREH

PAINAN

MUARA LABUH

MT. KERINTJI
BALAI SELASA

INDRAPURA
SUNGAI PENUH
LAKE KERINTJI

WEST COAST OF SUMATRA:
HEARTLANDS AND MARCHES

AFTER: SCHRIEKE, B., *INDONESIAN SOCIOLOGIGAL STUDIES*,
PART ONE, THE HAGUE and BANDUNG: van HOEVE,
1955.

HEARTLAND

The contrast we have in mind is . . . that between the predominantly [wet] rice producing areas and those where the cultivation of commercial crops is coming to the forefront; where, that is to say, commodity production takes place. In the latter areas money incomes are increasing by leaps and bounds compared with the former. But this contrast is not "remarkable" at all in reality; it is simply logical.[68]

Schrieke dates this "great change" from 1908–1912, when the last of the government-imposed restrictions on indigenous agriculture and trade were abolished, and when, more importantly,

---

[68] Schrieke, 1955.

the stimulus of rapidly expanding export markets began to be felt, at least in the marches. In the Kerintji highlands, the southern outlier, coffee exports rose from 190 tons in 1913, to 300 in 1923, to nearly 3,000 in 1926. In Muara Labuh, an isolated foothill region on the northern side of the mountain which gives Kerintji its name, children and even teachers stay away from school to work in the gardens, for "they see that a coffee farmer earns more than a miserable petty official, who is, moreover, obliged to work away from his native village." In Inderapura, also in the south, but on the coastal plain, "vast plantations can now be found where three years ago no coffee existed," and scarcely a garden is more than seven years old. Painan, in the western lowlands but farther north, "is still fairly backward, being a narrow and rather insignificant coastal region," but even here "the population has recently gone over to the cultivation of coffee." Coffee also flourishes northward in Lubuk Sikaping and eastward in Taluk; and "while in former days a large percentage of the population was driven to leave [these regions] for Malaya owing to shortage of food, many are returning now that gardens are being opened in their native [villages]."

In the lowland regions, especially those fringing the great eastern marsh, rubber is also of great, commonly even greater, importance, as for example in Bangkinang, where in 1925 a population of only 40,000 managed to export 1,250 tons of havea to Singapore, earning 180,000 pounds sterling. In two other lowland margin areas, one in the extreme north, the other in the extreme south, which twelve years earlier were so backward and impoverished that "the government had not dared to encourage rubber cultivation—in spite of the fact that . . . the population had the greatest difficulty paying its taxes, even when these were at a very low rate of assessment," there are nowadays people "to whom a rubber transaction running into some hundreds of pounds means nothing very extraordinary." Copra too is widespread, being most prominent around Pariaman on the northwest coast, where the growers not only process the coconuts themselves but carry the meat to market, thus eliminating the hated, expensive middleman. "Here we have to do," Schrieke sums up, "with a revolution in spirit similar to that of the early capitalist period in Europe as indicated by Max Weber and Sombart. The 'economic mentality' has made its entry upon the scene." [69]

This mentality has had its customary sociocultural accompaniments: increasing flexibility of land tenure; growth of individualism and slackening of extended family ties; greater class differentiation and conflict, intensified opposition between young and old, modern and conservative; weakening of traditional authority and wavering of traditional social standards; and even the growth of "Protestant ethic" religious ideologies.[70] What changed here (as in Java though in a different way) was not just a pattern of land use or a set of productive techniques but a system of functionally interrelated, adaptively relevant institutions, practices and ideas—a "cultural core."

The precise nature of the changes in the more technical aspects of this core is not fully spelled out by Schrieke; but the main development seems to have been a slowing down of the shifting cycle so as to lead to a transformation of the swidden pattern toward a more or less fixed gardening one. Rubber and coconut trees grown originally in or around swiddens eventually tend to become permanent or semipermanent groves if they are of commercial importance, because they are perennials, less easily choked by weeds, and less draining of the thin soil than are grains. Coffee is a greater problem because, more bush than tree, it provides less cover for the soil and has a shorter life span. But planted under specially selected shade trees on protected, or even terraced, highland slopes, coffee also can become an effective garden cultivation.[71] More than anything else it is rice,

---

[69] The preceding two paragraphs from Schrieke, 1955, esp., pp. 98–106.

[70] Schrieke, 1955, pp. 107 ff.

[71] Paerals, 1946. In the lowland areas coffee remained firmly enclosed within the swidden pattern, though even there it extended the life of a plot from two or three to seven or eight years, as well as leading to a

demanding light, encouraging weeds, and exhausting minerals, which makes shifting cultivation—at least in Southeast Asia— shift.

Thus, the reduction in the relative role of rice production as against tree crops has brought a movement away from true swidden toward permanent or semipermanent gardening. "Thinly populated" Muara Labuh, a onetime surplus swidden-rice producer, has turned into a heavy importer because "the inhabitants are too busily occupied cultivating coffee—to have time left for rice growing." In Inderapura, the population "cannot cope with [its] vast gardens owing to shortage of labor . . . [and consequently] tends to neglect rice growing—in favour of the cultivation of commercial crops." Rice fields are still being planted there, but under pressure exerted by the native administration, and it "is only achieved with difficulty." Rice exports from Kerintji have dropped nearly 75 percent in three years; and in Lubuk Sikaping "the cultivation of rubber and coffee has now put altogether an end to rice growing in forest clearings, the so-called 'shifting-cultivation.' " [72]

The reduction, or at least the limitation, of rice production in favor of commercial crops led of course to greater rice imports —in the first instance, from the plateau sawah areas, where this process was less in evidence than it was in the pioneering marches; in the second, from Java, which was having increasing difficulty merely feeding itself; and, finally, from surplus regions outside the archipelago (Burma, Thailand, Indochina). The process was cumulative. Where around 1922 a significant (within-Indonesia) rice export was still flowing from the west coast residency as a whole, most of it evidently from the heartland, by 1930 "the cultivation of coconuts, coffee and rubber had pushed rice into the background" to the point where the region

more careful temporal coordination of parallel swiddens. See Pelzer, 1945, pp. 25–26.
   [72] Schrieke, 1955.

was importing nearly 700 tons a year; by 1938, 3,300.[73] Thus, in two decades of an arm's-length encounter with European capitalism, the Minangkabau swidden peasant became what the Javanese sawah peasant, struggling to keep head above water for more than a century, never did: an acquisitive businessman fully enmeshed in a pecuniary nexus.

Again, the Minangkabau experience is perhaps not typical of Outer Indonesian development in all details. Each of the smallholder growing points was somewhat different culturally, socially, and ecologically, and so the exact pattern of their response to commercial opportunity was different. The heavy concentration on coconut gardens around Menado in northern Celebes (Minahasa) turned the region—which in the seventeenth and eighteenth centuries had been "the rice barn for the Moluccas"—into an inflexibly one-crop export economy at the full mercy of the copra market, because rice could not be grown in the shade of the long-lived coconut groves, and a decline in export returns thus brought almost immediate food shortages.[74] In the great rubber flats of Djambi, Palembang, and coastal Borneo, where the trees were more integrated into the established swidden complex, the situation was more flexible, for a decline in rubber prices could be at least partly compensated for by neglecting tapping for rice growing.[75] But even here the planting of rubber trees sometimes tended to transform subsistence peasants into export farmers and to crowd out food cultivation altogether, for example, at Benkalis along the Malacca

[73] Van de Koppel, 1946. Some of this rice went to feed laborers on Dutch concerns, for example the coal miners in the fields around Sawah Lunto. There were also some coffee, tea, and quinine estates, especially around Kerintji. In general, however, European enterprise in the west coast was limited.

[74] Van de Koppel, 1946.

[75] Van de Koppel, 1946. Around Deli much of the imported rice went to feed the imported Javanese and Chinese coolies. In 1931–1938 the residency of the east coast of Sumatra accounted for nearly half the rice imports into Outer Indonesia. Van de Koppel, 1946.

Strait where "rubber gardens occupied so large an area that there was actually a scarcity of land suitable for the making of new [swiddens] for food crops, and rubber trees had to be cut down." [76] In Atjeh, in Tapanuli, in the Lampungs, even in parts of the Lesser Sundas, such as Timor, the same picture of increased commercial exports and increased rice imports appears; while in the Moluccas, traditionally a sago-growing region, commercial gardening of coconuts permitted significant imports of rice into some areas.[77]

Thus, as the bulk of the Javanese peasants moved toward agricultural involution, shared poverty, social elasticity, and cultural vagueness, a small minority of the Outer Island peasants moved toward agricultural specialization, frank individualism, social conflict, and cultural rationalization. The second course was the more perilous, and to some minds it may seem both less defensible morally and less attractive aesthetically. But at least it did not foredoom the future.

---

[76] Pelzer, 1945, p. 25. Actually, scarcity of labor may have been as important a factor as scarcity of land; and, in any case, it evidently was the Dutch administration, not the peasants, who decided the trees "had" to be cut down.

[77] Van de Koppel, 1946; Ormeling, 1956. In a number of the Lesser Sundas, cattle and timber export played a minor role.

PART III

The Outcome

6. COMPARISONS AND PROSPECTS

The future came soon enough. The crash occurred in 1930: During the next four years, Netherlands East Indies export income dropped 70 percent.[1] In May, 1940, the Netherlands was occupied by the Germans, leaving the Indies in the unfamiliar role of a colony without a home country. By 1942 the vacuum was filled: its metropole was Tokyo. In August, 1945, the Indonesian Republic declared its Independence; by the end of 1949 it achieved it. Depression, war, occupation, and revolution— all in two decades.[2] About all that was left to happen was civil war and runaway inflation—and they came in the fifties.[3]

The Present Situation

But the effect of this series of convulsions on the basic pattern of the Indonesian economy has been less than one might expect. The transfer of sovereignty, the comprehensive politiciza-

---

[1] Van de Koppel, 1946.

[2] The definitive work on this period is Kahin, 1952. For economic developments during it, see Sutter, 1959, esp., Vols. I and II.

[3] On civil war, see van der Kroef, 1958; and Hanna, 1961, chapter vii. On inflation, see Higgins, 1957, pp. 13–25, 166–167.

tion of society, and the triumph of radical nationalism have profoundly altered the moral environment of economic activity. But for the general form of that activity Bergson's extravagantly historicist aphorism very nearly holds: there is nothing in the present but the past. The economy functions much less effectively, but (or, more exactly, because) it is the same economy. The threefold thematic structure announced by the Company, developed by the Culture System, and resolved by the Corporate Plantation System—technological dualism, regional imbalance, and ecological involution—persists; and the frustration of Indonesian aspirations persists with it.

Within the confines of this general framework, however, many things have changed. Petroleum aside, Europeans play a much reduced role in the capital intensive sector of the dual economy, and the Dutch, since the government takeover of their properties in December 1957, play virtually none at all.[4] The sugar industry, by far the hardest hit estate enterprise in the depression, much of its equipment reduced to scrap iron by the single-minded Japanese, and many of its mills burned to the ground by resentful Indonesian revolutionists, is but a shadow of its former self despite a recent upturn.[5] In Deli and environs hordes of peasant squatters—unemployed Javanese coolies, Sumatrans in-migrant from less fertile adjacent regions, local peasants freed at last from colonial regulations—have appropriated more than a hundred thousand hectares of former plantation land of all sorts for subsistence agriculture, a process started by Japanese attempts to make the area self-sufficient by setting plantation

---

[4] The mass takeover ostensibly stimulated by the West New Guinea (Irian) issue, affected more than 400 agricultural estates, as well as a number of banks, industrial firms, and transport lines. Paauw, in press.

[5] Sugar production fell from nearly 3,000,000 tons to 500,000 between 1930 and 1935, the only important estate crop to show a serious drop in output in the crisis. By 1940 it had recovered to about 1,500,000 tons, but in 1950 it was down to 275,000. Since then it has slowly risen to somewhat less than one-half the 1940 figures, one-quarter the 1930 figures. Van der Koppel, 1946; Statistical Pocketbook of Indonesia, p. 68.

laborers to work cultivating food crops on unused estates.[6] Between 1938 and 1955 the number of estates in exploitation in Indonesia just about halved; though in terms of actual planted area the decline was only slightly more than a quarter.[7]

At the same time the half-century drift toward industrial raw material exports at the expense of foodstuffs, spices, fibers, and the like had continued apace: between 1940 and 1956 the export volume of rubber, tin, and petroleum rose 40 percent while that of all other products fell by about a half; and the two-thirds of total 1940 export value quoted earlier for these three products had by 1956 become nearly three-quarters.[8] Thus the export sector of the economy narrowed down even more finely to Outer Indonesian—and especially Sumatran—products. Even the slight revival of sugar production in Java has done little to alter this picture, for the bulk of this product is now consumed domestically. If anything, regional imbalance has widened: in 1939 about 35 percent (by value) of Indonesian exports was shipped from Javanese ports, in 1956 only 17 percent.[9]

Involution, too, has proceeded relentlessly onward, or perhaps one should say outward, for a process which began to be felt first in full force mainly in the sugar regions is now found over almost the whole of Java. Against the background of our 1920 figure, those based on the 1955 election-registration (district analyses of the 1961 census figures are not yet available) show an assimilation of more and more of the island to the demographic—and, one presumes, the social—conditions prevailing in the densest ones during the earlier period (Graph 2). The mean

---

[6] Pelzer, 1957. A similar phenomenon, on a less extensive scale, has occurred on a number of the enclave-type estates in eastern Java.

[7] Statistical Pocketbook of Indonesia, pp. 64–65.

[8] Statistical Pocketbook of Indonesia, pp. 108, 129, 135.

[9] Statistical Pocketbook of Indonesia, p. 105. Actually as there was a significant amount of smuggling of small-holder copra and rubber from the Outer Islands in the latter year, the contrast was probably even greater; a point which obtains for the relative role of industrial raw materials in total export argument as well.

GRAPH 2

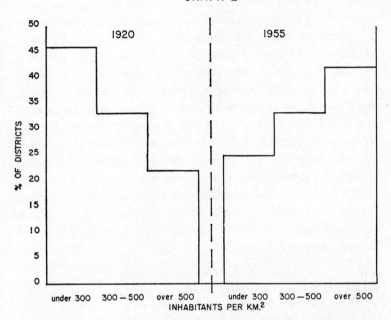

SOURCE: Calculated from Iso Reksohadiprodjo and Soedarsono Hadisapoe-tro, "Perubahan Kepadatan Penduduk dan Penghasilan Bahan Markanan (Padi) di Djawa dan Madura," Agricultura (Jogjakarta), I (1960), 3–107.

of this distribution of districts is about the same, but what was once the minority, the hyperpopulated district, is now the mode and what was once the mode, the comparatively moderately populated one, is now the minority. Analyzing these data more carefully, Iso Reksohadiprodjo and Soedarsono Hadisapoetro point out that, with a few exceptions, regions with heavy population remained in the same general density class from 1920 to 1930, but "regions with a low density of population in 1920 showed a tendency to a higher density in 1930 and still more markedly in 1955." [10] Thus, where the all-Java density rose by

---

[10] Iso Reksohadiprodjo and Soedarsono Hadisapoetro, 1960. A few areas with clearly adverse natural conditions, particularly with respect to irriga-

about 54 percent between 1920 and 1955, that of the main (former) sugar districts rose only by about 12, evidently—though no figures are available to prove it—as a result of greater urban migration from these depression-blighted areas, and perhaps in part of interregional migration to less populated rural areas.[11] The per-capita rice-production figures give the same picture. Such production fell by about a fifth in Java over these three and a half decades, but only by about a tenth in the sugar districts, partly because of the release of sawah from cane, which in some areas—notably the Sidoardjo delta south of Surabaja—actually managed to permit a maintenance of the 1920 per-capita paddy production.[12] In short, the sharpness of the line between "inner Inner Indonesia" and "Inner Indonesia" generally has been thoroughly blurred. What was once a local pathology is now an epidemic disease.[13]

---

tion, are an exception in that they have remained at a low level. Iso Reksohadiprodjo and Soedarsono Hadisapoetro, 1960.

[11] Calculated from footnote 10 and Landbouwatlas. As the 1955 figures are only for Indonesian citizens and those for 1920 presumably for all residents, the totals are not precisely comparable, but the error thus introduced is minor. Also only 73 of the original 98 sugar districts can be used due to lacunae in the 1955 data and to administrative reorganizations between the two dates. The 25 omitted districts are, however, well scattered, and their inclusion is not likely to change the relationships significantly.

[12] Calculated from footnote 10. The Brantas-delta regencies of Sidardjo and Modjokerto, nearly a third of whose sawah was in cane in 1920, had a per-capita production of about 450 kg in that year, about 453 in 1955, in the face of a 10 percent population increase. Footnote 10, and Wertheim, 1956, p. 371. The generally more favorable ecological conditions in the former sugar areas, permitting terraces to be more thoroughly squeezed, also played a role; but these figures are only for rice, and the areas would probably be pretty well offset by more dry land (tegal) cultivation in the nonsugar districts, were the data available.

[13] Another sign that involution is now entering its acute phase is that fragmentation of ownership seems also to be spreading now that virtually no more land is available to bring into cultivation and the island is extremely crowded almost everywhere. In Malang regency, Bennet (1961) found that the average size of sawah holding dropped 9 percent between

Thus, although political, economic, and intellectual disorder has reached—at least on the national level—a stage of near catastrophe, ecologically speaking the postwar picture in Indonesia is about the same as before the war, perhaps more so. Most of Java is crowded with post-traditional wet-rice peasant villages: large, dense, vague, dispirited communities—the raw material of a rural, nonindustrialized mass society. In the rest of Inner Indonesia—Bali, Lombok, and some of the less involuted areas of Java—as well as in some Outer Indonesian pockets (the Minangkabau heartland, the Tapanuli Batak, southwest Celebes), traditional wet-rice peasant villages predominate, where density, though high, is not yet overwhelming, and the established kinship, political, and religious institutions remain powerful. Outer Indonesia is characterized by localized plantation and mining enclaves, where capital-intensive techniques are mixed with an uprooted labor force; by the small-holder export crop gardeners —now seriously hampered by government economic policies— clustered in certain areas; and by the traditional, mainly kinship-organized swidden cultivations spread over the rainforest interiors of the larger western islands and the rugged monsoon forest hills of the smaller eastern islands.[14]

1942 and 1953, of tegal holdings 13 percent, while population climbed about 11 percent, "possibly indicating that increasing densities of population are being reflected here less in an absence of sawah holding and more in a diminishing size of the average holding." He estimates that although 85 percent of the households have some land, only about 64 percent of them have holdings of a size adequate to their subsistence needs.

[14] This typology, intentionally gross and ignoring urban settlements altogether, is based on the more discriminating classification of Indonesian communities in H. Geertz, in press. There are, also, as a result of both pre- and post-revolutionary government efforts, some Javanese, Sundanese, and Balinese "transmigrated" wet-rice settlements in the Lampungs at the southern tip of Sumatra, and to a much lesser extent in southwest Celebes. Between 1937 (when there were already some 60,000 Inner Indonesian settlers in the Lampongs) and 1956, about 286,000 people either moved or were (voluntarily) moved from Java to these areas. Robequain, 1954, p. 162; Statistical Pocketbook of Indonesia, p. 17. For a description

In human history "final" pictures are final until they change. But, despite the sloganeering about "winds of change," "the awakening East," or "the revolution of rising expectations," and despite, too, the real possibility of a totalitarian triumph in Djakarta, there is no evidence that the major outlines of the Indonesian pattern of adaptation, of the plurality of diverse cultural cores which compose it, are likely to alter in the foreseeable future. For all the proclaimed, if only half-believed, optimism at its apex, Indonesia at its base is an anthology of missed opportunities, a conservatory of squandered possibilities.

Java and Japan

A sharper sense for these lost opportunities and possibilities is developed if one looks to another area where they were seized—not allowed to slip through the fingers like the proverbial chances of a lifetime. Historically, "what might have happened" of course did not; and from one point of view there is little else to be said about it. But the fact that it might have is not without significance, for "there are moments in history the whole record of which is not contained in the story of what happened, and the meaning of which is not disclosed except in an estimate of what never came to be." [15] It is, as Frost has written, roads not taken which make all the difference.

For Java, the obvious comparative case is Japan. Much differs between them: geography, history, culture, and, of course, per-capita income—Japan's being about twice Java's.[16] But much, too, is similar. Both are heavily populated. Both rest agriculturally on a labor-intensive, small-farm, multicrop cultivation regime centering on wet rice. Both have managed to maintain

---

of transmigration efforts in the Colonial Period, see Pelzer, 1945; in the Republican period, Utomo, 1957.

[15] Brinkley, 1935, p. 299.

[16] Estimated gross national product per capita (1955) for Japan is about $240, for Indonesia about $127. Ginsburg, 1961, p. 18. Javanese per-capita product is, of course, below that of Indonesia as a whole, though by how much it is impossible to say.

a significant degree of social and cultural traditionalism in the face of profound encounter with the West and extensive domestic change. In fact, in agriculture, the further back one goes toward the mid-nineteenth century the more the two resemble one another. Japanese per-hectare rice yields at the beginning of Meiji (1868) were probably about the same as those of Java at the beginning of the Corporate Plantation System (1870); today they are about two and a half times as high.[17] Between 1878 and 1942 the percentage of the Japanese labor force employed in agriculture dropped from around 80 to around 40; the Javanese figure for the end of the nineteenth century is not known, but in 1930—and probably still today—it had not fallen below 65 percent.[18] And, though it is even more a matter of head-long estimate, the percentage of aggregate net income contributed by agricultural production in the Japan of the 1880's was of the same general order as that in the Java of the 1950's, by which time the Japanese percentage was only a third as large.[19] Given, then, all the admittedly important background differences, one can hardly forbear to ask when one looks at these two societies: "What has happened in the one which did not happen in the other?"

A satisfactory answer to such a question would involve the whole economic, political and cultural history of the two civilizations; but even if we confine ourselves to predominantly ecological considerations, a number of dramatic differences leap to the eye. The most striking—and the most decisive—is the contrast between the way Japan utilized its rapid population increase and the way Java utilized hers. Between 1870 and 1940 Java absorbed the bulk of her increase in numbers—about thirty million people

---

[17] Ohkawa and Rosovsky, 1960; FAO Yearbook, pp. 36–37.

[18] Ohkawa and Rosovsky, 1960; Statistical Abstract for the Netherlands East Indies, 1935, pp. 143–146.

[19] Ohkawa and Rosovsky, 1960. Mears, 1961. The Javanese figure is again estimated on the basis of the all-Indonesia figure, the percentage of income from agriculture being probably lower in Java than in the Outer Islands.

—into post-traditional village social systems of the sort already described, but

in the first century of modernization, Japan maintained a relatively unchanging population in agriculture while the total population increased two and one-half fold. Practically all the increase in the labor force was absorbed in non-agricultural activities. There was little change in the size of the rural population. Almost all the natural increase of the national population was absorbed in urban areas.[20]

From 1872 to 1940, the Japanese farm population remained virtually constant around 14 million people (or 5½ million households) at the same time as the total population grew approximately 35 million.[21] Comparable statistics on the farm population of Java are not available, but that it grew at better than an average of one percent a year during this entire period seems a conservative estimate.[22] Japan, in short, did not involute; which should cut the ground out from under any charges of "paddy-field determinism" which might be brought against our Javanese analysis. But what, then, did it do?

For one thing, it increased agricultural productivity *per worker,* not just *per terrace.* "Using the latest and best computations," Ohkawa and Rosovsky estimate that the productivity of agricultural labor (net output/labor force in agriculture) increased 2.6 percent annually between 1878 and 1917.[23] Thus, as I have attempted (with much less adequate computations) to

---

[20] Taeuber, 1960.

[21] Namiki, 1960; Ohkawa and Rosovsky, 1960.

[22] Starting around 1920, Javanese towns seem to have begun, evidently for the first time, to grow more rapidly than the general population, and this trend has accelerated since the revolution (Wertheim, 1956, pp. 185–186; The Siau Giap, 1959). However even today they come nowhere near absorbing the entire increase as the Japanese towns seem to have done since the last quarter of the nineteenth century.

[23] Ohkawa and Rosovsky, 1960. Toward the end of the period this rate tended to slacken, and after the first world war it dropped to about 1 percent. But by that time Japan's industrial sector had been firmly launched.

demonstrate, where Java increased per-hectare yields at least to the First World War but not per worker yields, Japan increased both over roughly the same period. The contrast is all the more impressive, because, in Japan as in Java, the basic structure of proprietary control, the general form of the producing unit, and the over-all pattern of rural culture seem to have been relatively unaltered:

The increases in output and productivity were based on the traditional patterns of rural organization inherited, in the main, from the Tokugawa period. The small family farm, averaging about 1 hectare per household, the distribution between peasant proprietors and tenants, high rents in kind—all of these characteristics were maintained during the [1878–1917] period. At the same time, there was no strong trend of land consolidation and this preserved the scattered holdings of tiny plots of ground.[24]

No move to extensify agriculture; no marked trend toward a class polarization of large landlord and rural proletarian; no radical reorganization of the family-based productive unit—characteristics of Java and Japan alike since the turn of the century.[25] But Japan increased productivity per agricultural worker 236 percent, Java—the estates aside for the moment—hardly increased it at all.[26] For Java, *plus ça change, plus c'est la même chose* may be a fitting epitome. But for Japan it would have to read *plus c'est la même chose, plus ça change.*

The readiest explanation for this difference—and the one most commonly invoked—is the greater technological advance in Japanese agriculture. Irrigation was expanded, land reclaimed, seed

---

[24] Ohkawa and Rosovsky, 1960. For a description of the still fairly traditional quality of village life in Japan—in many ways more traditional than in Java—see Beardsley, Hall and Ward, 1959.

[25] As the fundamental groundwork for the Javanese type of pattern was laid in the Culture System period, so that for the Japanese was laid in the equally decisive Tokugawa (1600–1868), particularly the later phases of it. On this, see Smith, 1959.

[26] Ohkawa and Rosovsky, 1960.

selection improved and fertilizer use increased, cooperative activity became more effective and widespread, planting was intensified, weeding and harvesting methods developed, agricultural knowledge increased.[27] Though the contrast between the two technologies is often exaggerated, on one level this thesis is beyond question—the sharp rise in productivity per farmer in Japan clearly must rest on "key improvements in Japanese agricultural practice in keeping with the small unit of production." [28] Yet it is, nonetheless, unsatisfactory. First, it involves a somewhat uncritical reading of the present into the past. Until about the turn of the century, by which time Japan had irrevocably "taken off" and Java had definitively involuted, the difference between either the rate or level of technological advance in the two peasant agricultures was not so great as it has been since then.[29] Second, even to the extent the argument is factually valid, it merely restates the question: it is just this difference in technological progress that we wish to explain. There is little in

---

[27] Ohkawa and Rosovsky, 1960; Dore, 1960. Again, as in Java, these changes began well back in the previous (here the Tokugawa) period. See Smith, 1959, pp. 87–107.

[28] Ohkawa and Rosovsky, 1960.

[29] Rostow tentatively dates Japanese take-off in the 1878–1900 period, 1960, p. 38. A full documentation of the argument that Javanese technological advance in agriculture very nearly kept pace with the Japanese until the beginning of this century or the end of the last would require a thorough review of the two technological histories, which—particularly as the Javanese one has yet to be written—cannot be attempted here. It is worth pointing out, however, that all main improvements listed for pre-twentieth century Japanese agriculture, with the partial exception of artificial fertilization, seem to have been present in Java. Also, it is of note in this connection that in 1910–1911, when the first reliable comparative figures appear, Japanese per-hectare rice yields are only 80 percent higher than Javanese, while a decade later they are 300 percent higher (Wickizer and Bennett, 1941, p. 318), the difference evidently stemming from radically rising inputs of (now mostly commercial) fertilizer, which increased more than 650 percent between 1898–1902 and 1913–1917 (Ohkawa and Rosovsky, 1960).

the two technologies around, say, 1870—the end alike of the Culture System and of the Tokugawa period, which could account for their divergence since that time.

More genuinely determinative of the separation into contrasting courses was the manner in which a traditional labor-intensive, Lilliputian, family-farm, wet-rice-and-second-crop type of ecosystem came to be related to a set of modern economic institutions. Specifically, where Japanese peasant agriculture came to be complementarily related to an expanding manufacturing system in indigenous hands, Javanese peasant agriculture came to be complementarily related to an expanding agro-industrial structure under foreign management. As labor productivity in the capital-intensive sector in Japan increased, it increased also in the labor-intensive sector; as it increased in the capital-intensive sector in Java it remained approximately constant in the labor-intensive one. In Japan, the peasant sector supported the industrial one during the crucial three decades of the latter's emergence largely by means of extremely heavy land taxation; in Java, the peasant sector supported the industrial one through the provision of underpriced labor and land. In Japan, the industrial sector, once underway, then re-invigorated the peasant sector through the provision of cheap commercial fertilizer, more effective farm tools, support of technical education and extension work and, eventually, after the First World War, simple mechanization, as well as by offering expanded markets for agricultural products of all sorts; in Java most of the invigorating effect of the flourishing agro-industrial sector was exercised upon Holland, and its impact upon the peasant sector was, as we have seen, enervating. The dynamic interaction between the two sectors which kept Japan moving and ultimately pushed her over the hump to sustained growth was absent in Java. Japan had and maintained, but Java had and lost, an integrated economy.

To a great extent, Japan maintained it and Java lost it in the critical four decades of the mid-nineteenth century—1830–1870.

At the same time that van den Bosch was superimposing an export-crop economy upon the traditional Javanese sawah system, Japan had locked herself away from Western interference and was moving toward a more commercialized, less immobile rural economy on its own.[30] In both societies, peasant agriculture was becoming, within a generally unchanging basic pattern, steadily more labor intensive, more skillful and more productive. But in Java the increase in output was soon swamped by the attendant spurt in population; in Japan the population remained virtually constant.[31] In both societies, the peasant's agricultural productivity increased. But in one it was, so to speak, reserved (largely through the operation of a tributational tax system) for, as it turned out, future investment in an indigenous manufacturing system. In the other, it was immediately expended to subsidize the swelling part-time labor force (i.e., the peasant in his two-fifths corvee role) of a foreign-run plantation system, and its potential for financing a properly Indonesian take-off dissipated.

[30] Smith, 1959. This movement went on, to some extent, during the whole Tokugawa period, though it seems to have come to a climax in the last century of it, and was further speeded up after Perry's visit in 1853, at which time the seclusion policy was, at least officially, ended. In any case, steady, if slow and locally uneven improvement in Javanese agriculture might also be traced well back into the seventeenth century were the material available. It is possible that Japan "started lower" than Java, its natural conditions being less suitable for rice, so that its growth in productivity between the beginning of the seventeenth century and the end of the nineteenth would have had to have been greater in order for it to achieve Javanese levels by the later date. See Rosovsky, 1961, p. 81, note 104.

[31] Japanese population seems to have increased more or less steadily from about the end of the twelfth century to the beginning of the eighteenth, at which time it stabilized, evidently because the traditional ecosystem had found its climax equilibrium. Between 1726 and 1852 population was virtually constant. After 1852 and contact with the West it began to rise, but slowly and irregularly. After 1870 rapid rise began, moving from about 35 to 55 million in less than half a century (1873–1918). Taeuber, 1958, pp. 20–25, 44–45.

But to comprehend this tale of two economies fully requires also a comparison of the denouements, which means extending the four-decade critical period to its nine-decade consummation in take-off into sustained growth on the one hand and in involution into static expansion on the other. In outline form, the contrasting patterns of development in the two societies between approximately 1830 and the end of the First World War can be summarized as follows:

	JAVA	JAPAN
TECHNIQUE	Gradual improvement through the entire period (and in all likelihood before), but in a wholly labor-intensive manner.	Gradual improvement through the entire period (and, in fact, the whole Tokugawa period). This also took place mainly in a labor-intensive manner until around 1900, after which rapidly increasing capital inputs, mostly in the form of fertilizer, took place.
POPULATION	Rapid growth after 1830, evidently as a result of declining mortality due to improved communications and greater security and of increased fertility due to the labor-tax pressures of the Culture System; but not, save possibly for a brief initial period, as a result of generally rising Indonesian living standards.	Rapid growth began only after 1870, evidently as a result of a decline in the death rate attendant on a rising national standard of living and of increasing fertility due (indirectly) to expanding employment opportunities in manufacturing.[32]
EMPLOYMENT	No significant expansion outside of traditional agricultural pursuits, but a rapid expansion within them made possible by the perfection of labor-absorbing productive techniques which raised land but not labor productivity. Peasants provided unskilled occasional labor for (first government, then private) plantations, but at a price well below its marginal pro-	Rapid expansion in the industrial sector, absorbing the whole population increase. Agricultural employment virtually constant, both land and labor productivity rising, the latter about 70 percent more rapidly.[33]

	JAVA	JAPAN
	ductivity, the costs of their subsistence being largely borne by the village economy.	
URBANIZATION	Retarded. Towns and cities grew much less rapidly than total population, and the depressive effect upon fertility rates commonly associated with urban life were largely absent, delaying the usual postindustrial slowing of population growth.[34]	Accelerated, particularly after the Restoration. Towns and cities grew much more rapidly than total population. As rates of natural increase were moderately depressed in the urban areas, this was in the main due to a jump in rural-urban migration.[35]
PER-CAPITA INCOME	Probably close to constant over the whole period in the peasant sector; rapidly rising in the plantation sector.	Rising with increasing rapidity in the peasant sector, the rise being used to finance an even more rapid rise in the manufacturing sector after Meiji.[36]
ECONOMIC DUALISM	Increasingly severe. Increased capital inputs into the plantation sector, increased labor inputs into the peasant. Separation between the two sectors cultural, social, and technological at the same time, with little intermediate industrial activity.	Marked, but moderated by close cultural, social, and economic connections between the two sectors and by the flowering of small-scale industrial activity.[37]

[32] The diachronic comparison of the two demographic developments is perhaps most simply expressed in terms of the changing ratio of the Japanese to Javanese population: 1830, 3.8; 1870, 2.1; 1900, 1.6; 1920, 1.6; 1955, 1.6 (calculated from Taeuber, 1958, pp. 22, 46, 70; and from sources given in note 42, chapter iv, above). Thus, the early rise of Japanese population gave it somewhere around 3½ to 4 times the Javanese total by the beginning of the nineteenth century, a gap the Javanese explosion nearly cut in half by 1870, after which time the convergence slowed as the two populations grew after 1900 at about the same annual rate—between 1 and 1½ percent.

[33] Ohkawa and Rosovsky, 1960.

[34] "Development in [Southeast] Asia, centered as it was on plantations, mines, oil fields and exports of raw materials, brought more *industrialization* than *urbanization*; the checks on family size brought by the urban industrialization and the New World operated less effectively in the

Whatever value this comparison of Java and Japan may have does not lie in any assumption that had Java been "left alone" she would have followed the Japanese path, or even that she would now be in an economically more viable state. What would have happened had the Dutch not colonized the Indies is clearly not even an hypothetically answerable question, for it depends upon what historical events would have occurred instead, and their number is infinite. Nor does the value of the comparison rest on the assumption that Java's course now must be to re-enact somehow the Japanese pattern if take-off is to be at length accomplished. The world has moved on, both in and outside Java, and the alternatives which face her today are not those which faced Japan a century ago. Its value lies in providing a contrasting, yet comparable case which can shed light on what

---

underdeveloped countries. The drops in fertility rates came eventually in most Asian countries too, but too late to prevent serious population pressure from arising before planned economic development began." Higgins, 1958.

[35] Taeuber, 1960. Here, it was not so much the fact that urbanization showed an initially higher over-all growth, but rather its greater development prevented Japanese transition rates from climbing as high between 1880 and 1920 or so as Javanese ones seem to have climbed between 1840 and 1880.

[36] "Being fearful of the political consequences of foreign borrowing, the [Meiji] government financed investment almost entirely from domestic sources—mainly agriculture. The land tax accounted for 78% of ordinary revenues (the bulk of total revenues) from 1868 to 1881, and although the figure tended to fall after that it still stood at 50% in 1890. High as the rate of tax on land was, however, it did not represent an increase over the Tokugawa period. Already at the end of that period the take from agriculture by the warrior class was immense, and the Meiji government merely redirected it into new channels. Modernization was achieved, therefore, without reducing rural living standards or even taking the increase in productivity that occurred." Smith, 1959, p. 211. As this view seems to neglect increased efficiency in tax collection it may be slightly optimistic so far as pressure on the peasants is concerned.

[37] On the importance of Japanese small industry, see Rosovsky and Ohkawa, 1960.

happened in Java and therefore on the nature of her present situation. The economic history of Japan is not a norm from which Java has, alas, departed, nor that of Java a pathology from which Japan has, praise God, escaped. Rather Japanese economic history is Javanese with a few crucial parameters changed (a proposition which could be stated equally well the other way around), and in this consists its comparative significance.

There are two major parameters which are so changed: the existence of colonial government in Java is replaced in Japan by the existence of a powerful indigenous elite; the development of capital-intensive agriculture in Java is replaced in Japan by the development of a capital-intensive manufacturing system. Behind these major parametric differences lie a host of others. The strength of the Japanese elite grew out of the traditional, religiously supported patterns of political loyalty characteristic of the culture generally;[38] Java's colonization was in part a mere reflex of her geographical location at the cross-roads of the Orient, of her neighbors' possession of the right spices at the wrong time, and of the inherent fragility of the classical Indonesian states. The tropical climate and other physical characteristics of Java which, as noted, made export sugar and subsistence rice natural dualistic partners were lacking in Japan, whose more temperate conditions perhaps made a plantation farming pattern less adaptive. Other differences—in the world views of both masses and elites, in micro-ecological conditions, in pre-seventeenth-century historical development, in patterns of social stratification and mobility, in market systems, and so on—could also be cited in the same connection. A full comparative analysis would have to trace them out (or at least the more powerful of them—for the list of parameters whose change might affect the behavior of any given system has no end) and attempt to assess their relative weight.[39]

---

[38] On Japanese concepts of loyalty (or "obligation") see Benedict, 1946; on their religious basis, Bellah, 1957.

[39] For an incisive theoretical discussion of the role of the concept of parameter in system analysis, see Ashby, 1960, esp. pp. 71–79.

But on the ecological, and to an extent the economic, level their expression was funneled through these two most immediately decisive differences. It is the predicament of all science that it lives by simplification and withers from simplicisticism.

The existence of colonial government was decisive because it meant that the growth potential inherent in the traditional Javanese economy—"the excess labor on the land and the reserves of productivity in the land," to use a phrase which has been applied to the "slack" in the Japanese traditional economy at the Restoration—was harnessed not to Javanese (or Indonesian) development but to Dutch.[40] This is not a mere matter of monetary returns (though, leaving subsistence farming aside, in 1939 assessed per-capita income in the European community in Indonesia was more than a hundred times that in the Indonesian community),[41] for certainly the growth of the plantation industry made possible greatly increased "native welfare" expenditures—on health, rural credit, and so on—in Java. Nor is it a mere matter of immediate benefit or harm, for some accompaniments of its growth certainly redounded, in a residual way, to the short-run advantage of the peasantry—better irrigation, improved communications, increased availability of foreign manufactures, and the like. Fundamentally, it is a matter of the transformative impact upon society implicit in modern industry. The improvement of human capital and the expansion of physical capital; the creation of a modern business class and the crystallization of an efficient market system; the formation of a skilled and dis-

---

[40] The quotation is from Ranis, 1959; cited in Rosovsky, 1961.

[41] Calculated from Kahin, 1952, p. 36. The figures are income-tax statistics, and as Indonesian wage incomes under 900 guilders were not assessed, they exaggerate the contrast somewhat. On the other hand, they included Outer Island Indonesian "commercial" incomes, and, as we have seen, these were significantly higher than Javanese. Finally, the 1939 population is estimated, the last colonial census having occurred in 1930. On balance, the times-a-hundred figure is probably conservative. An estimate including Indonesian subsistence production put the 1939 ratio at about 60 to 1. Polak, 1942, p. 60.

ciplined work force and the raising of labor productivity; the stimulation of higher propensities to save and the construction of workable financial institutions; the inculcation of an entrepreneurial outlook and the development of more effective forms of economic organization—all these to a significant extent endogenously generated cultural, social, and psychological resources upon which industrialism feeds were in a sense exported with the commodities the plantations produced. The difference in "economic mentality" between Dutch and Javanese which Boeke took to be the cause of dualism was in fact in great part its result. The Javanese did not become impoverished because they were "static"; they became "static" because they were impoverished.

The fact that the form in which capital-intensive industry came to Java was agricultural simply reinforced this process. Much more than manufacturing, industrial agriculture—and especially sugar cultivation—permits a sharp division of labor between a traditionalized labor force and a modernized managerial elite. The Japanese peasant had to go to town and become a full-time, reasonably disciplined member of a manufacturing system, even if the organization of his factory was modeled along traditional lines and his ties with his village homeland were kept green to ease the transition.[42] The Javanese peasant did not, literally, even have to move from his rice terrace. Plantation agriculture is a much more effective way of marrying a nonindustrial labor force to an industrial productive apparatus than is manufacturing, whose functional requirements are inevitably more stringent. No matter how strongly traditional elements are maintained in manufacturing, they inevitably result in some serious dislocations in life-ways and some major reorientations in outlook for those caught up in them at all levels. They are a common school for modernism, as Japanese history since 1920, and especially since 1945, demonstrates. Whether the blunting of

---

[42] On the persistence of traditional social and cultural forms in modern industrial settings in Japan, see Abegglen, 1958.

such effects is an essential characteristic of plantation industry or a merely accidental one is perhaps debatable, and it is always all too easy to assume that the way things worked out in fact is the way things had to work out in principle. One might at least conceive of an agro-industrial system which is as effective a school for mass economic modernization as manufacturing, and in Hawaii, for example, one might actually find an approximation of one. But it can hardly be gainsaid that such a system has a strong natural bias toward the production of what Mintz has called a rural proletariat—a hapless coolie labor force which achieves the agonies attendant upon industrialization without achieving its cultural, social, and psychological fruits.[43] The real tragedy of colonial history in Java after 1830 is not that the peasantry suffered. It suffered much worse elsewhere, and, if one surveys the miseries of the submerged classes of the nineteenth century generally, it may even seem to have gotten off relatively lightly. The tragedy is that it suffered for nothing.

The Outline of the Future

Yet, though much has been lost, much remains. Indonesia's per-capita food consumption may now finally be falling after its long period of relative constancy, but it is still not so low as it is in India, Morocco, Tanganyika, or Burma—in fact, in South and Southeast Asia it is significantly exceeded probably only by Malaya and the Philippines.[44] Nor is there a

---

[43] Mintz, 1956.

[44] See Ginsburg, 1961, p. 30, for estimated per-capita per-day caloric intake for 93 countries, among which Indonesia stands 73d at about 2,040. In 1957, the "ideal" (i.e., satisfactory) *carbohydrate* consumption per person per year was set by the Nutrition Institute in Djakarta at about 160 kg (standardized in terms of the caloric content of rice), in which year it was in fact an estimated 159 in Java, 170 in the Outer Islands, 162 over-all. Even without food imports it was about 155, 159, and 155. Calculated from Mears, Afiff and Wreksoatmodjo, 1958. The Moluccas, where data are inadequate, have been omitted. As only rice, maize, cassava, and sweet potatoes were used in the calculation, it is conservative, 160 kg of

radically uneven distribution of national wealth—Indonesia probably has less of a plutocratic problem than any other country in Asia save possibly China. And, even more important, for all its century-and-a-half drift toward economic paralysis Indonesia's economy still has some "slack" left in it, some exploitable potentialities for growth. It is discouraging that these seem in the process of being squandered too.

Aside from petroleum and other mining enterprises, and what manufacturing exists, the main possibilities of investing capital in agricultural economy would seem to be three. First, there is the one major technical advance in Japanese farming which has not as yet been fully applied in Java—fertilization, and in association with it, improved seed selection.[45] Second, there is the swidden-to-commercial-gardening "agricultural revolution" in some parts of Outer Indonesia. And third, there is the old estate sector, now mostly under Indonesian state management. But capitalizing on these various opportunities is more difficult than it often looks to agronomists, economists, or nationalist politicians.

The fact that fertilization (and improved seed selection) could significantly increase Indonesia's, including even Java's, farm yields seems well established.[46] In seven hundred and fifty field trials in Java it was found that "any suitable [chemical] fertilization gives a crop response of at least 20%, in many cases 30%, and

---

rice per person per year amounting to approximately 1,600 calories per person per day. Even aside from the intrinsic unreliability of the data, caloric intake is, of course, a very gross measure of food supply, but the Indonesian diet is probably not yet radically deficient in vitamins and minerals either.

[45] A certain amount of sawah fertilization took place, largely as a secondary result of sugar-growing, in the last years of the colonial period, and American aid programs provided some $50 million worth of fertilizer in the first years following independence. How much scientifically based rice-seed improvement there has been is less clear.

[46] For a general review, see Mears, 1957.

in some cases much higher." [47] Even allowing for the possibility that some of these increases may prove less spectacular and less lasting in practice than they have in trials, the potential for increasing Indonesia's food supply through artificial fertilizers can hardly be doubted. But, problems of manufacture, costs, and so on aside, a merely technological view is inadequate. No more than the earlier improvements in irrigation, cultivation techniques, and crop diversification can an increase in yields through fertilization be viewed as a way out of Java's—and Indonesia's—agrarian dilemma. The figures in Table 4 with respect to the expansion of nonrice cultivation indicate more than a 20 percent rise in agricultural productivity over a few decades, but nothing ultimately came of that in the way of increases in per-capita income, much less sustained growth. It has been estimated that the Dutch spent 250 million guilders on the construction of irrigation and flood-control facilities between about 1880 and 1939, but the Javanese peasant at the end of it was just where he had been at the beginning.[48] The central problem is not whether Java can still increase her output: the central problem is whether, as in Japan, this increase (perhaps at long last the traditional terrace's final spurt) can be captured to finance an industrial sector into which the increasing population can be absorbed. Otherwise the economy is still treading water, which is a dubious improvement over drowning outright unless help eventually arrives.

Since the revolution, a rising proportion of the population increase has been absorbed—"collected" would be a better word—into the towns as redundant civil servants, time-passing students, underemployed traders, and all but workless unskilled laborers. Thus Indonesia is moving from industrialization without urban-

---

[47] Hauser, Georg, an FAO agricultural chemist and adviser to the Institute of Soil Research in Bogor, as quoted in Mears, Afiff and Wreksoatmodjo, 1958.

[48] Irrigation investment estimates from Pelzer, 1945.

ization toward urbanization without industrialization.[49] And so far as nonagricultural production fails to rise, the economic effect is much the same as if the population shunted off to the towns was being taken directly into the village system (where, in fact, the bulk of the increase is still being absorbed), for the share-the-poverty pattern is simply extended to embrace both rural and urban dwellers. A rise in agricultural productivity would certainly alleviate this for a time; but it would only postpone a day of reckoning probably already too long put off. Only if most of the rise can be siphoned off into industrial investment (including investment in "human capital") does it seem that fertilizers and the rest will really contribute something dynamic to the general economic situation. If it is not so siphoned off, it will merely further accelerate the process of involution, not just in agriculture now, but throughout the whole society.

It is, thus, an ironic and melancholy fact that so far as fertilizer application is, in truth, Java's last trump in peasant agriculture, it is being played at a time when the general political and cultural atmosphere is such that it almost certainly is being wasted, in much the same way as all her earlier trumps were wasted. In the absence of any genuine reconstruction of Indonesian civilization, despite the rhetoric of revolutionary populism, any alteration of the persisting direction of its development, pouring fertilizer onto Java's Lilliputian fields is likely, as modern irrigation, labor-intensive cultivation and crop diversification before it, to make only one thing grow: paralysis. This is not an argument against the technical improvement of peasant farming. It is not an argument against the openness of human history. It is an argument against a myopic pragmatic optimism which

---

[49] Not that there has been *no* industrialization. But where total population seems to have grown about 10 percent between 1940 and 1952, "total industrial employment" seems to have grown only about seven. A comparable figure for urban population growth is unavailable, but as the urban population more than doubled between 1930 and 1952, it would almost certainly be much higher. Paauw, in press; The Population of Indonesia.

allows short-run gains to obscure the general trend of events, which isolates purely technical improvements from the historically created cultural, social, and psychological context in which they are set, and which, because of these failings, exacerbates the ailments it sets out to cure.

The potentialities inherent in the Outer Island swidden-to-export garden development are equally real and, for about the same reasons, equally problematical. That small scale-commercial farming of industrial crops can be a powerful source of economic dynamism in Indonesia is amply proved by the prewar experience, when, for example, small-holder rubber acreage jumped more than 300 percent in a decade and a half (1925–1940) and "brought forth a stream of money the like of which was unknown in the Indies." [50] But enthusiasm is tempered by the mere fact that Outer Indonesia is, after all, part of Indonesia as a whole. The involutional process in Java, even more than world price movements or technological change, is the most immediate constituent of the economic landscape within which such farming takes place. That such a landscape is an unpromising one for growth ought to be apparent by now.

The whole problem can best be seen in terms of rubber, increasingly the preponderant small-holder export crop.[51] To say small-holder export farmer today is largely to say rubber grower (or, to a secondary extent, coffee or copra grower), and so the question of the power of Outer Island agriculture to jolt Indonesia's economy out of its deepening rut largely comes down to the question of whether rubber can become a leading sector in a national transition to sustained growth.

Obviously, a wide range of special factors, stretching from the

[50] The statistic is from Bauer, 1948, p. 16; the quotation is from van Gelderen, 1961.

[51] With 1938 as 100, the small-holder production of rubber was 327 in 1954; coffee 68, copra 54, pepper 23, average of all small-holder export production excluding rubber, 48. In value terms, returns from small-holder rubber were about 60 percent more than from coffee, copra, and pepper combined. Higgins, 1957, pp. 149, 155 (calculated).

competition of synthetics to the changing input needs of world industry, is relevant to any such assessment, to say nothing of international political conditions. (Small-holder rubber production reached its highest point in history during the Korean war.[52]) But, in terms of the analysis being pursued here, they all come to an immediate focus in two main considerations: the degree to which rubber-farming is a more profitable use of Outer Island land and labor than subsistence farming, and the degree to which the difference can be channeled into growth investment, either in the extension and improvement of the rubber industry itself, or in other productive enterprises. In these terms, the Javanese involutional process acts against development: as it progresses it converts more of the export income earned by rubber etc. into imported foodstuffs and other consumption necessities (textiles, etc.), and so gradually incorporates rubber into itself, as it has already incorporated irrigation improvements and crop diversification, and threatens to incorporate fertilization. The cultivation and export of rubber become just one more way in which per-capita income can be maintained at a nearly constant level as population growth continues at a relatively unchanged rate.[53] Outer Island export farming becomes a mere tail to the kite of Javanese subsistence farming; rather than pulling it in new directions, it is merely dragged along after it; rather than jolting, it stabilizes it.

No more than in fertilization, irrigation, or crop diversification, however, is this an inevitable outcome, an intrinsic product

---

[52] Thomas, K. D., n.d., Appendix A, Table I. Domestic marketing arrangements are also of central importance, of course; see Thomas, n.d., pp. 45–58.

[53] In 1939, the import value of the four most important consumption necessities (rice, wheat flour, fish, and textiles) was about 8 percent of the total import value, about 5 percent of the total export value. For 1950–1956 it was 15 percent of the import value, 12 percent of the export value. Calculated from Statistical Pocketbook of Indonesia, pp. 111, 99. By 1959, rice alone had reached nearly 20 percent of the total import value. Paauw, in press.

of some mysterious developmental force, but merely, in the present state of Indonesian national life, a likely outcome, and in fact is already occurring. In particular, it has been a combination of deficit financing—including subsidization of rice imports— and an unrealistically "favorable" official exchange rate which has been the mechanism through which an increasing proportion of the energizing force of the Outer Indonesian "agricultural revolution" has been converted into Inner Indonesian "static expansion." [54] The result of such policies (and others) has been to arrest the swidden-to-garden transition in mid-passage. The rising local price of rice and other consumption goods relative to that of rubber, the inability of small holders to gain control of the capital to invest in processing equipment, planting material, labor and so forth, and the general disincentive quality attached to a sense that the bulk of locally earned revenue is being utilized extralocally, have combined to discourage not so much production as, more importantly, any improvement in productivity.[55]

The prewar "agricultural revolution" of the rubber areas was in truth but a semirevolution:

To [capitalize on the potential value of vast stretches of available land extremely suitable for the growing of rubber trees] it was not necessary to break with the old system of cultivation. On the abandonment of an exhausted [swidden], rubber-tree seeds were left

---

[54] For a review of Indonesian fiscal policies until 1956, see Higgins, 1957, pp. 1–36.

[55] From 1950–1955 small-holder rubber production remained more or less constant (Higgins, 1957, p. 149); since then it may have declined somewhat, but due to the problem of smuggling the exact picture is not clear. The Population of Indonesia authors estimate 1950–1954 small-holder per-hectare yields of rubber at a constant 4.2 (as against 5.8 for the somewhat bedeviled estates), but this is the grossest of guesses. There is a difference in quality as well: small-holder production averaged about 65 percent of total output during 1951–1955, but small-holder value only 58 percent of total value. Thomas, K. D., n.d., Appendix A, Table I, and Appendix E, note.

in the ground. After some years the result was a growth of trees ready for tapping. In this way the [swidden] system itself produced groves over a series of years culminating around 1910, as a result of which rapid increase in production was possible after 1916. Moreover, there was more systematic planting of rubber trees independent of food crops, particularly in Tapanuli and in regions outside Sumatra. But even in such cases one can hardly speak of a higher form of production. The owners of the rubber groves merely applied the *separo* ["halves"] system traditionally used for food crops. Under it the worker was rewarded with half of whatever yield he had produced. A characteristic trait of the adaptation is that the owner buys the product from his workers at the current local price. Hence his gross income is the same as that of all his workers together. Their wages are determined directly by the market price of the product. As long as that is high enough to provide a level of living sufficiently inducing for them to go on tapping, production continues. The lower limit is thus that at which half of the production provides the workers with the minimum level of subsistence acceptable. Other costs are negligible. Practically nothing is done towards the systematic maintenance of the groves, the construction of processing equipment, and the like. The most expensive piece of machinery is a hand mangle for pressing slabs. On the other hand, in the present situation in the Indies the upper limit for Native production is not determined by the natural limit, that is to say the number of tappable trees and their production capacity, but by the amount of labour available. There are still extensive areas waiting to be tapped. The link with food production has not been broken . . . the people can constantly fall back on their food-crop fields.[56]

As in the Javanese one-foot-in-the-terrace,-one-in-the-mill pattern, this arrangement was commonly praised by colonial scholars and officials (a) for its elasticity, in that it could expand and contract with changing market conditions, thus protecting the peasant against the vagaries of international commodity markets, and (b) for its adaptability, in that in terms of given

---

[56] Van Gelderen, 1961.

factor proportions (high land, low labor, low capital) it was the most efficient use of resources.[57] But, as we have seen, the maximization of adaptability is a spiritless goal, elasticity perpetuated becomes flaccidity, and a highly efficient combination of resources in terms of "given" production coefficients may be a simple alternative to the more genuinely entrepreneurial activity of changing the coefficients. If small-holder rubber (or some other commercial crop) is to be a leading sector in take-off, it will have to be pursued with a fullness of commitment—and an acceptance of attendant risk—which will make its sustained development possible. It will have to become a growing industry, not a functional equivalent, in a habitat which cannot support wet rice, for that most highly perfected device for running faster while staying in the same place, the Javanese sawah. The "extensive areas waiting to be tapped" will hardly prove much of a dynamic force when they *are* tapped if labor productivity about matches "the minimum level of subsistence acceptable," and population (now probably increasing at better than 2 percent a year in the Outer Islands)[58] and aggregate income once more advance *pari passu.*

The completion of the swidden-to-garden revolution, whether in rubber, copra, or coffee, demands increased investments in planting materials, improved labor methods, more careful processing, and a more effective distributive network. That this is not occurring seems apparent in the face of conversion of an increasing percentage of foreign exchange into imported subsistence goods and a narrowing of the price differential between rubber and rice.[59] Small-holder cultivation of export crops re-

---

[57] Van Gelderen, 1961.

[58] Sumaniwata, 1962.

[59] As all figures for rubber are export and not production figures and since no detailed research has been undertaken into small-holder rubber-growing, it is difficult to pin down what is happening at the ecological level. But it seems clear that planting, even under the most favorable estimates, is not keeping pace with even the replacement demands im-

mains a source of succor for the hard-pressed Indonesian economy, but it is hardly even approaching the status of a driving force.

The prospects for the third possible source of dynamism in the agriculture sector, the plantations, seem, at the moment, even less promising. The fact that the estates were in foreign hands until recently, has made their effective functioning in an atmosphere of ardent nationalism extremely difficult. Squatters, insecurity, labor difficulties, the prospective lapsing and nonrenewal of leases, and the ever-present threat of expropriation (a threat which has now been realized so far as the Dutch are concerned) do not add up to an environment conducive to foreign investment.[60] The tobacco estates have been hampered —mainly by squatters—to the point where plantation tobacco production in 1955 was about a fifth that of 1939, and continues to fall.[61] Tea was similarly hard hit, production falling 70 percent in the same period.[62] In fact, of all the estate crops, only rubber showed an increase—about 35 percent—and, again, the rise in output gives a deceptively optimistic picture: to maintain

---

plicit in current production so that "most of the acreage is over 25 years old" (and much is over 35), as does the fact that planting materials remain poor. Quality assessment likewise is an elusive matter, but there is also little doubt that small-holder quality remains low because of undeveloped technique even in terms of present technology (which is itself unchanging). For a balanced view of the technical side of small-holder rubber-growing, hampered only by the absence of reliable data, see Thomas, K. D., n.d., *passim*. Allen and Donnithorne (1957, p. 137) estimate that about a fifth of Indonesian small-holder rubber acreage is now "practically worthless" due to age and overtapping, so that "high production by the peasants has been accompanied . . . by a very large consumption of capital invested in rubber."

[60] For a description of the political context within which the estates operated during the first five years of independence, see Sutter, 1959, III, pp. 695–766. For 1955–59, see Hanna, 1961, chapter iv.

[61] Statistical Pocketbook of Indonesia, p. 61.

[62] Statistical Pocketbook of Indonesia, p. 61.

productivity the estates estimate they should have replanted 160,000 hectares between 1945 and 1955; they seem to have actually replanted about 35,000.[63] The result is "that Indonesia, which in the thirties had been ahead of other countries in technique has now fallen behind Malaya in the use of up-to-date-methods" and "the danger is that Indonesia may be unable to compete, not merely with synthetic rubber but with natural rubber produced by other regions which have invested new capital in high-grade planting material and up-to-date processing plants." [64] Over-all, estate export production in 1955 was about three-quarters of what it had been in 1938; if rubber, living evidently on borrowed time, is excluded, the figure drops to less than half.[65] Estate agriculture hardly seems like the leading edge of anything right now, unless it be collapse.

With the 1957 expropriation of the Dutch plantations, about half of the total estate output has come under direct governmental control, the other half remaining, for the while, in trembling foreign hands.[66] Whether this nationalization, that at long last reduced the racial element in dualism and brought the capital-intensive sector more firmly within the boundaries of the Indonesian economy proper, is a prelude to a revival of the estate sector remains to be seen. But the shortage of trained managerial personnel, the relentless spread of governmental immobilism, and the flourishing of ideological extremism born of a general failure of nerve makes it most unlikely. "The economic situation of the state is deteriorating," said President Sukarno, who certainly ought to know, in mid-1959. "The financial situation of the state is deteriorating, the social condition of society is

---

[63] Thomas, K. D., n.d., p. 17.

[64] Allen and Donnithorne, 1957, pp. 136, 137. Indonesia which became the world's largest rubber producer for a while after World War II is now again second to Malaya, her production having fallen about 7.5 percent since the Korean war boom of 1951, "at a time when world production, demand and prices are rising." Hanna, 1961, chapter iv.

[65] Higgins, 1957, p. 149.

[66] Paauw, in press.

deteriorating—in all fields we are deteriorating." [67] A search for the true diagnosis of the Indonesian malaise, takes one, thus, far beyond the analysis of ecological and economic processes to an investigation into the nation's political, social, and cultural dynamics.

---

[67] Quoted in Hanna, 1961, chapter ii.

BIBLIOGRAPHY

BIBLIOGRAPHY

Abegglen, J.
  1958 The Japanese Factory. Glencoe, Illinois, The Free Press.
Adam, L.
  n.d. Enkele Gegevens Omtrent den Economische Toestand
        van den Kalurahan Sidoardjo. Weltevreden, Kolff.
Allen, G. C., and A. G. Donnithorne
  1957 Western Enterprise in Indonesia and Malaya. New York,
        Macmillan.
van Alphen, H.
  1870 Java en het Kultuurstelsel. s'Gravenhage, van Stockum.
Ashby, E.
  1960 Design for a Brain. New York, John Wiley (2d ed.).
Bastin, J.
  1957 The Native Policies of Sir Stamford Raffles in Java and
        Sumatra. Oxford, Oxford University Press.
Bates, M.
  1952 Where Winter Never Comes. New York, Scribners.
  1953 Human Ecology. In Kroeber, 1953, pp. 700–713.
Bauer, P.
  1948 Report on a Visit to the Rubber Growing Smallholdings
        of Malaya, July–September, 1946. London, Her Majesty's
        Stationery Office.

Beardsley, R. K., J. W. Hall and R. E. Ward
1959   Village Japan. Chicago, University of Chicago Press.
Beets, A. N. J.
1946   Vorstenlandse Tabak. *In* van Hall and van de Koppel,
        1946, IIB, 414–486.
Bellah, R.
1957   Tokugawa Religion. Glencoe, Illinois, The Free Press.
Benedict, R.
1946   The Chrysanthemum and the Sword. Boston, Houghton
        Mifflin.
Bennet, D.
1961   Three Measurements of Population Pressure in Eastern
        Java. Ekonomi dan Keuangan, 14:97–106.
Bernet Kempers, A. J.
1959   Ancient Indonesian Art. Cambridge, Harvard University
        Press.
Beukering, J. A. van
1947   Het Ladangvraagstuk, een Bidrijfs- en Sociaal Econo-
        mische Probleem. Batavia, Mededeelingen v.h. Departe-
        ment v. Economische Zaken in Nederlandsch-Indie, No. 9.
Die Bie, H. C. H.
1902   De Landbouw der Inlandsche Bevolking op Java. Batavia,
        Kolff.
Boeke, J. H.
1910   Tropische-Koloniale Stadhuishoudkunde. Amsterdam, dis-
        sertation.
1947   The Evolution of the Netherlands Indies Economy.
        Haarlem, H. D. Tjeenk Willink.
1953   Economics and Economic Policy of Dual Societies. Haar-
        lem, H. D. Tjeenk Willink.
Brinkley, R. C.
1935   Realism and Nationalism. New York, London, Harper.
Broek, P. J. van den
1946   Bevolkingstabak. *In* van Hall and van de Koppel, 1946,
        IIB, 522–558.
Burger, D. H.
1928   Rapport over de Desa Pekalongan in 1869 en 1928.
        Weltevreden, Kolff (Economische Bischrijvingen).

1930 Vergelijking van den Economischen Toestand der Districten Tajoe en Dijakenan. Weltevreden, Kolff (Economische Bischrijvingen).

1939 De Ontsluiting van Java's Binnenland voor het Wereldverkeer. Wageningen, H. Veenman.

Clarke, G.

1954 Elements of Ecology. New York, John Wiley.

Collins, W. B.

1959 The Perpetual Forest. Philadelphia, New York, Lippincott.

Conklin, H.

1954 An Ethnoecological Approach to Shifting Agriculture. Transactions of the New York Academy of Sciences, Series II, 17:133–142.

1957 Hanunoo Agriculture in the Philippines. Rome, Food and Agricultural Organization of the United Nations.

1959 Shifting Cultivation and the Succession to Grassland. Proceedings, 9th Pacific Science Congress (1957), 7:60–62.

1960 The Cultural Significance of Land Resources among the Hanunoos. Philadelphia Anthropological Society Bulletin, 13:38–42.

Day, C.

1904 The Dutch in Java. New York, Macmillan.

Dice, L. R.

1955 Man's Nature and Nature's Man. Ann Arbor, University of Michigan Press.

Dobby, E. H. G.

1954 Southeast Asia. London, University of London Press (4th ed.).

van Doorn, C. L.

1926 Schets van de Economische Ontwikkeling der Afdeeling Poerworedjo (Kedu). Weltevreden, Kolff.

Dore, R. P.

1960 Agricultural Improvement in Japan: 1870–1900. Economic Development and Cultural Change, 9 (part ii): 69–91.

Encyclopædia Britannica (Tobacco). Eleventh Edition, Cambridge, Cambridge University Press, 1911.

Encyclopedia van Nederlandsche–Indie. Leiden, Brill, 1899–1905.

FAO Yearbook of Food and Agricultural Statistics 1955.

Firey, W.
  1947 Land Use in Central Boston. Cambridge, Harvard University Press.

Forde, C. D.
  1948 Habitat, Economy and Society. New York, Dutton.

Freeman, J. D.
  1955 Iban Agriculture. London, Her Majesty's Stationery Office.

Furnivall, J. S.
  1944 Netherlands India. Cambridge, Cambridge University Press.
  1948 Colonial Policy and Practice: A Comparative Study. Cambridge, Cambridge University Press.

Geddes, W. R.
  1954 The Land Dayaks of Sarawak. London, Her Majesty's Stationery Office.

Geertz, C.
  1956 Religious Belief and Economic Behavior in a Central Javanese Town: Some Preliminary Considerations. Economic Development and Cultural Change, 4:134–158.
  1959 The Javanese Village. In Skinner, 1959, pp. 34–41.
  1960 The Religion of Java. Glencoe, The Free Press.

Geertz, H.
  In press Indonesian Cultures and Social Structure. In McVey, in press.

van Gelder, A.
  1946 Bevolkingsrubbercultuure. In van Hall and van de Koppel, 1946, III:427–75.

van Gelderen, J.
  1929 Western Enterprise and the Density of the Population in the Netherlands Indies. In Schrieke, 1929, pp. 85–102.
  1961 The Economics of the Tropical Colony. Indonesian Economics, 1961, pp. 111–164.

Ginsburg, N.
  1961 Atlas of Economic Development. Chicago, University of Chicago Press.

Glamann, K.
1958 Dutch Asiatic Trade, 1620–1740. The Hague, Nijhoff;
Copenhagen, Danish Science Press.
Goethals, P. R.
1961 Aspects of Local Government in a Sumbawan Village.
Ithaca, Cornell University Press, Modern Indonesia Project Monograph Series.
Goffman, E.
1961 Asylums. New York, Anchor Books.
Goldenweiser, A.
1936 Loose Ends of a Theory on the Individual Pattern and
Involution in Primitive Society. *In* Lowie, 1936, pp. 99–
104.
Gonggrijp, G.
1957 Schets Ener Economische Geschiedenis van Indonesie.
Haarlem, Bohn (4th printing).
Gourou, P.
1953a L'Asie. Paris, Hachette.
1953b The Tropical World (trans., E. D. Laborde). New York,
Longmans Green.
1956 The Quality of Land Use of Tropical Cultivators. *In*
W. L. Thomas, 1956, pp. 336–349.
Graaf, E. A. van de
1955 De Statistiek in Indonesië. s'Gravenhage, van Hoeve.
de Graaf, H. J.
1949 Geschiedenis van Indonesie. s'Gravenhage and Bandung,
van Hoeve.
Grist, D. H.
1959 Rice. London, Longmans Green (3d ed.).
Haeckel, E.
1870 Ueber Entwickelungsgang und Aufgabe der Zoologie.
Jenaische Zeitschrift für Medicin und Naturwissenschaft,
5:353–70.
van Hall, C. J. J.
1946 Bevolkingsthee. *In* van Hall and van de Koppel, 1946,
IIB, 246–271.
n.d. Insulinde, De Inheemsche Landouw. Deventer, van Hoeve.

van Hall, C. J. J., and C. van de Koppel (eds.)
1946   De Landbouw in den Indischen Archipel. s'Gravenhage, van Hoeve, 4 vols.

Halpern, J. M.
1961   The Economies of Lao and Serb Peasants: a Contrast in Cultural Values. Southwestern Journal of Anthropology, 17:165–77.

Hanna, W. A.
1961   Bung Karno's Indonesia. American Universities Field Staff.

Hawley, A.
1950   Human Ecology. New York, Ronald Press.

Higgins, B.
1956   The 'Dualistic Theory' of Underdeveloped Areas. Economic Development and Cultural Change, 4:99–115.
1957   Indonesia's Economic Stabilization and Development. New York, Institute of Pacific Relations.
1958   Western Enterprise and the Economic Development of Southeast Asia: a Review Article. Pacific Affairs, 31:74–87.
1959   Economic Development. New York, Norton.

Hollinger, W.
1953a   Indonesia, Quantitative Studies 2, The Food Crop Sector. Center for International Studies, Massachusetts Institute of Technology (dittoed).
1953b   Indonesia, Quantitative Studies 3, Tables on the Food Crop Sector. Center for International Studies, Massachusetts Institute of Technology (dittoed).

Honig, P., and F. Verdoorn (eds.)
1945   Science and Scientists in the Netherlands Indies. New York, Board for the Netherlands Indies, Surinam and Curacao.

Hose, C., and W. MacDougal
1912   The Pagan Tribes of Borneo. London, Macmillan.

Huender, W.
1921   Overzicht van den Economischen Toestand der Bevolking van Java en Madoera. s'Gravenhage, Nijhoff.

Huntington, E.
1945   Mainsprings of Civilization. New York, John Wiley.

Indonesia
  1956 Human Relations Area File, Subcontractors Monograph,
        No. 57. New Haven.
Indonesian Economics. The Hague, van Hoeve, 1961.
Iso Reksohadiprodjo and Soedersono Hadisapoetro.
  1960 Perubahan Kepadatan Penduduk dan Penghasilan Bahan
        Makanan (Padi) di Djawa dan Madura. Agricultura
        (Jogjakarta), 1:3–107.
Jay, R.
  1956 Local Government in Rural Central Java. Far Eastern
        Quarterly, 15:215–27.
Kahin, G. McT.
  1952 Nationalism and Revolution in Indonesia. Ithaca, Cornell
        University Press.
Kampto Utomo
  1957 Masjarakat Transmigran Spontan Didaerah W. Sekam-
        pung (Lampung). Djakarta, P. T. Penerbitan Universitas.
van Klaveren, J. J.
  1955 The Dutch Colonial System in Indonesia. Rotterdam (?),
        no publisher.
Koens, A. J.
  1946 Knolgewassen. In van Hall and van de Koppel, 1946, IIA,
        163–240.
Kloff, G. H. van der
  1929 European Influence on Native Agriculture. In Schrieke,
        1929, pp. 103–125.
  1937 The Historical Development of Labor Relationships in
        a Remote Corner of Java as They Apply to the Cultiva-
        tion of Rice. New York, no publisher.
  1953 An Economic Case Study: Sugar and Welfare in Java. In
        Ruopp, 1953, pp. 188–206.
Koningsberger, V. T.
  1946 De Europese Suikerrietcultuur. In van Hall and van de
        Koppel, 1946, IIA:278–404.
Koppel, C. van de
  1946 Eenige Statistische Gegevens over de Landbouw in
        Nederlandsch-Indie. In van Hall and van de Koppel,
        1946, I:361–423.

Kroeber, A. L.
1939 Cultural and Natural Areas of Native North America. Berkeley, University of California Press.
1953 (ed.), Anthropology Today. Chicago, University of Chicago Press.

Kroef, J. M. van der
1958 Disunited Indonesia. Far Eastern Survey. 27:49–63, 73–80.
1960 Land Tenure and Social Structure in Rural Java. Rural Sociology. 25:414–30.

Kuperus, G.
1930 De Bevolkingscapaciteit van de Agrarische Bestaansruimte in de Inheemsche Sfeer up Java en Madoera. (Omstreeks, 1930.) Nederlandsch Aardrijks Kundig Genootschap. LXI:363–409.
1938 The Relation Between Density of Population and Utilization of Soil in Java. Comptes Rendus du Congres International de Geographie, II (section iii):465–77.

Laan, P. A. van der
1946 Deli Tabak. *In* van Hall and van de Koppel, 1946, IIB, 353–415.

Landbouwatlas van Java en Madoera. Mededeelingen van het Central Kantoor voor de Statistiek, no. 33. s'Gravenhage, Nijhoff, 1926.

Leach, E.
1954 Political Systems of Highland Burma. Cambridge, Harvard University Press.
1959 Hydraulic Society in Ceylon. Past and Present, 15:2–25.

Lekkerkerker, C.
1916 Land en Volk van Sumatra. Leiden, Brill.
1938 Land en Volk van Java. Batavia, Wolters.

Levy, M.
1952 The Structure of Society. Princeton, Princeton University Press.

Living Conditions of Plantation Workers and Peasants on Java in 1939–40 (trans., R. van Neil). Ithaca, Modern Indonesia Project, Translation series, Cornell University, 1956.

Lowie, R. (ed.)
1936 Essays in Anthropology Presented to A. L. Kroeber. Berkeley, University of California Press.

Maas, J. G. J. A., and F. T. Bokma
1946   Rubbercultuur der Ondernemingen. *In* van Hall and van
          de Koppel, 1946, III, 235–426.
McVey, R. (ed.)
In press   Indonesia. Hew Haven, Human Relations Area Files.
Matsuo, T.
1955   Rice Culture in Japan. Tokyo, Yokendo.
Mears, L. A.
1957   The Use of Fertilization as One Means of Reducing the
          Problems Associated with the Distribution of Rice in
          Indonesia. Ekonomi dan Keuangan. 10:570–80.
1961   Economic Development in Indonesia Through 1958.
          Ekonomi dan Keuangan, Indonesia, 14:15–57.
Mears, L. A., S. Afiff and H. Wreksoatmodjo
1958   Rice Marketing in Indonesia, 1957–58. Ekonomi dan
          Keuangan, 11:530–570.
Metcalf, J.
1952   The Agricultural Economy of Indonesia. Monograph 15,
          U.S. Department of Agriculture, Washington, D.C.
Meulen, W. A. van der
1949–50   Irrigation in the Netherlands East Indies. Bulletin of
          the Colonial Institute of Amsterdam, 3:142–159.
Mintz, S.
1956   Cañamelar: the Sub-culture of a Rural Sugar Plantation
          Proletariat. *In* Steward, 1956.
1958   Labor and Sugar in Puerto Rico and in Jamaica, 1800–
          1850. Comparative Studies in Society and History. 1:273–
          283.
Mohr, E.
1945   The Relation Between Soil and Population Density in
          the Netherlands Indies. *In* Honig and Verdoorn, 1946,
          pp. 254–262.
1946   Bodem. *In* van Hall and van de Koppel, 1946, I:9–62.
Murphey, R.
1957   The Ruin of Ancient Ceylon. Journal of Asian Studies,
          XVI:181–200.
Namiki, M.
1960   The Farm Population in the National Economy Before

and After World War II. Economic Development and
Cultural Change, 9 (part ii):29–39.
Odum, E. P.
   1959   Fundamentals of Ecology. Philadelphia and London,
          Saunders.
Ohkawa, K., and H. Rosovsky
   1960   The Role of Agriculture in Modern Japanese Economic
          Development. Economic Development and Cultural
          Change, 9 (part ii):43–67.
Ormeling, F. J.
   1956   The Timor Problem. Groningen, Djakarta and s'Graven-
          hage, Wolters and Nijhoff.
Ottow, S. J.
   1957   De Oorspong der Conservatieve Richting. Utrecht, Oos-
          thoek.
Paauw, D.
   In press   The Indonesian Economy. In McVey, in press.
Paerels, B.
   1946   Bevolkingskoffiecultuur. In van Hall and van de Koppel,
          1946, IIB, 89–119.
Park, R. E.
   1934   Human Ecology. American Journal of Sociology, 42:1–15.
   1936   Succession as an Ecological Concept. American Sociologi-
          cal Review, 1:171–179.
Pelzer, K. J.
   1945   Pioneer Settlement in the Asiatic Tropics. New York, In-
          stitute of Pacific Relations.
   1957   The Agrarian Conflict in East Sumatra. Pacific Affairs,
          30:151–59.
Pendleton, R. L.
   1947   The Formation, Development and Utilization of the Soils
          of Bangkok Plain. Natural History Bulletin 14.
Polak, J. J.
   1942   The National Income of the Netherlands Indies. New
          York, Institute of Pacific Relations.
The Population of Indonesia. Ekonomi dan Keuangan Indonesia.
   9:1–27, 1956.
Quinn, J. A.
   1950   Human Ecology. New York, Prentice-Hall.

Raffles, T. S.
  1830   The History of Java. London, John Murray, 2 vols.
Raka, I. Gusti Gde
  1955   Monografi Pulau Bali. Djakarta, Pusat Djawatan Pertanian Rakjat.
Ranis, G.
  1959   The Financing of Japanese Economic Development. Economic History Review, Vol. II, No. 3.
Reinsma, R.
  1955   Het Verval van het Cultuurstelsel. s'Gravenhage, van Veulen.
Reyne, A.
  1946   De Cocospalm. *In* van Hall and van de Koppel, 1946, IIA, 427–525.
Richards, P. W.
  1952   The Tropical Rain Forest. Cambridge, Cambridge University Press.
de Ridder, J.
  n.d.   De Invloed van de Westersche Cultures op de Auctochtone Bevolking van der Oostkust van Sumatra. Wageningen, H. Veenman.
Robequain, C.
  1954   Malaya, Indonesia, Borneo and the Philippines (trans., E. D. Laborde). New York, Longmans Green.
Rosovsky, H.
  1961   Capital Formation in Japan. New York, The Free Press.
Rosovsky, H., and K. Ohkawa
  1960   The Indigenous Components in the Modern Japanese Economy. *In* Essays in the Quantitative Study of Economic Growth, Economic Development and Cultural Change. 9:476–501.
Rostow, W. W.
  1960   The Stages of Economic Growth. Cambridge, Cambridge University Press.
Ruopp, P.
  1953   Approaches to Community Development. The Hague and Bandung, van Hoeve.
Rutgers, J. L.
  1946   Peper. *In* van Hall and van de Koppel, IIB, 620–654.

Scheltema, A. M. P. A.
1930   De Sawahoccupatie op Java en Madoera in 1928 en 1888.
       Korte Mededeelingen van het Central Kantoor voor de
       Statistiek. Buitenzorg.
1931   Deelbouw in Nederlandsche-Indie. Wageningen, Veen-
       man.
1936   The Food and Consumption of the Native Inhabitants
       of Java and Madura. Batavia, Institute of Pacific Relations.
Schrieke, B.
1929   (ed.) The Effect of Western Influence on Native Civiliza-
       tions in the Malay Archipelago. Batavia, Kolff.
1955   Indonesian Sociological Studies, Part I. The Hague and
       Bandung, van Hoeve.
1957   Indonesian Sociological Studies, Part II. The Hague and
       Bandung, van Hoeve.
Sears, P. B.
1939   Life and Environment. New York, Columbia University
       Press.
Semple, E. C.
1911   Influences of Geographic Environment. New York, Holt.
Skinner, G. W. (ed.)
1959   Local, Ethnic and National Loyalties in Village Indo-
       nesia. New Haven, Yale University Cultural Report
       Series, Southeast Asia Studies.
Smith, T.
1959   The Agrarian Origins of Modern Japan. Stanford, Stan-
       ford University Press.
Spate, O. H. K.
1945   The Burmese Village. The Geographical Review, Vol.
       XXV.
Statistical Abstract for the Netherlands East Indies, 1935. Batavia,
       Landsdrukkerij.
Statistical Pocketbook of Indonesia, 1957. Djakarta, Biro Pusat Sta-
       tistik.
van Steenis, G.
1935   Maleische Vegetatieschetsen. Tijdschrift v.d. Nederland-
       sche Aardijkskundig Genootschap, pp. 25–67, 171–203,
       363–398.

Steward, J.
   1955   Theory of Culture Change. Urbana, University of Illinois Press.
   1956   (ed.) The People of Puerto Rico. Urbana, University of Illinois Press.
Sumaniwata, S.
   1962   Sensus Penduduk Republik Indonesia, 1961 (preliminary report). Djakarta, Central Bureau of Statistics.
Sutter, J. O.
   1959   Indonesianisasi. Data Paper No. 31–1. Southeast Asia Program, Cornell University. Ithaca, 4 vols.
Taeuber, I.
   1958   The Population of Japan. Princeton, Princeton University Press.
   1960   Urbanization and Population Change in the Development of Modern Japan. Economic Development and Cultural Change, 9 (part ii):1–28.
Tergast, G. C. W.
   1950   Improving the Economic Foundation of Peasant Agriculture on Java and Madoera (Indonesia). Unpublished mimeographed paper, trans. by the International Bank for Reconstruction and Development.
Terra, G. J. A.
   1946   Tuinbouw. In van Hall and van de Koppel, 1946, IIA, 622–746.
   1958   Farm Systems in South-East Asia. Netherlands Journal of Agricultural Science, 6:157–181.
The Siauw Giap
   1959   Urbanisatie Problemen in Indonesie. Bijdragen tot de Taal-, Land- en Volkenkunde, 115:249–276.
Thomas, K. D.
   n.d.   Smallholders Rubber in Indonesia. Djakarta, Institute for Economic and Social Research, University of Indonesia (mimeographed).
Thomas, Jr., W. R. (ed.)
   1956   Man's Role in Changing the Face of the Earth. Chicago, University of Chicago Press.

Thompson, W. S.
    1946   Population and Peace in the Pacific. Chicago, University of Chicago Press.
Veer, van der
    1946   Rijst. *In* van Hall and van de Koppel, 1946, IIA, 7-110.
Veth, P. J.
    1912   Java, Geographisch, Ethnologisch, Historisch. Haarlem, Bohn (2d ed.).
Vollenhoven, C.
    1906   Het Adatrecht van Nederlandsch-Indie. Leiden, Brill.
de Vries, H. M.
    n.d.   The Importance of Java Seen from the Air. Batavia, Kolff.
Wertheim, W. F.
    1956   Indonesia in Transition. The Hague, van Hoeve.
Wertheim, W. F., and The Siauw Giap
    1962   Social Change in Java, 1900–1930. Pacific Affairs, 35:223–247.
Wickizer, V. D., and M. K. Bennett
    1941   The Rice Economy of Monsoon Asia. Stanford, Stanford University Press.
Wissler, C.
    1926   The Relation of Man to Nature in Aboriginal America. New York, Oxford University Press.
Wittfogel, K.
    1957   Oriental Despotism. New Haven, Yale University Press.

INDEX

INDEX